The Institute of Biology's
Studies in Biology no. 150

Monotremes and Marsupials: the Other Mammals

Terence J. Dawson
Professor and Head, School of Zoology,
University of New South Wales,
Kensington, New South Wales, Australia

Edward Arnold

First published 1983
by Edward Arnold (Publishers) Limited
41 Bedford Square, London WC1 3DQ

British Library Cataloguing in Publication Data

Dawson, Terence J.
Monotremes and marsupials.—(Institute of Biology studies in biology ISSN 0537–9024; 146)
1. Monotremata
I. Title II. Series
599.1 QL737.M1

ISBN 0–7131–2853–4

Printed and bound in Great Britain at
The Camelot Press Ltd, Southampton

General Preface to the Series

Because it is no longer possible for one textbook to cover the whole field of biology while remaining sufficiently up to date, the Institute of Biology proposed this series so that teachers and students can learn about significant developments. The enthusiastic acceptance of 'Studies in Biology' shows that the books are providing authoritative views of biological topics.

The features of the series include the attention given to methods, the selected list of books for further reading and, wherever possible, suggestions for practical work.

Readers' comments will be welcomed by the Education Officer of the Institute.

1983

Institute of Biology
20 Queensbury Place
South Kensington
London SW7 2DZ

Preface

Monotremes such as the platypus and marsupials like koalas and kangaroos, are regarded with great interest and affection by people at large. Within the scientific community however, monotremes and marsupials have tended to have a 'bad press'. Generally they have been considered to be primitive and conservative, and not really proper mammals like the placental mammals. However, in the last two decades in Australia, these animals have begun to be appreciated as the successful inhabitants of the earth that they are. Of necessity, in this book I have had to concentrate on those areas of monotreme and marsupial biology, such as reproduction, temperature and energy relations and brain function, in which these mammals were once regarded as being particularly primitive. To give this discussion perspective, comparisons often have been made with placentals, and I hope the end result is to give a better appreciation of the true diversity of mammalian abilities. In this endeavour I have been helped by Dominic Fanning who gave real substance to my roughly drafted figures, and my wife Dr. Lyndall Dawson who helped eliminate much of the scientific jargon from the text.

Sydney 1983 T.J.D.

Contents

General Preface to the Series iii

Preface iii

1 Origins of Mammals and the Relationships of Monotremes 1
1.1 Origins of mammals 1.2 Relationships of monotremes to other mammals 1.3 Relationships within the monotremes

2 General Features and Natural History of Monotremes 9
2.1 The platypus, *Ornithorhynchus anatinus* 2.2 The echidna or spiny anteater, *Tachyglossus aculeatus* 2.3 Long beaked echidnas, *Zaglossus bruijni*

3 Reproductive Biology of Monotremes 13
3.1 Monotreme eggs 3.2 Female reproductive anatomy 3.3 Oviparity in monotremes: an insight into viviparity? 3.4 Breeding cycles 3.5 Lactation and mammary glands 3.6 Reproduction in male monotremes 3.7 Poison spur and crural system

4 Energy and Temperature Relations of Monotremes 22
4.1 The characteristics of homeothermy 4.2 Historical attitudes to homeothermy in monotremes 4.3 Metabolic relationships of monotremes 4.4 Temperature regulation in the cold 4.5 Hibernation 4.6 Temperature regulation at high temperatures

5 Cardiovascular and Neural Physiology of Monotremes 28
5.1 Cardiovascular physiology 5.2 Brain and intelligence

6 Marsupials: Origins and Historical Biogeography 32
6.1 The tribosphenic molar 6.2 Site of origin of marsupials 6.3 The Tertiary radiation of marsupials 6.4 Marsupial extinctions: Pleistocene and Recent

7 Marsupial Types, Their Habits and Relationships 39
7.1 Problems of marsupial classification 7.2 Marsupial families 7.3 Evolutionary relationships of marsupials

8 Reproduction in Marsupials 51
8.1 Reproductive anatomy of female marsupials 8.2 Reproductive cycles of marsupials 8.3 Foetal development and placental relationships 8.4 Relative efficiency of marsupial reproduction 8.5 Evolution of viviparity

9 Energy and Temperature Relationships in Marsupials 64
9.1 Historical attitudes 9.2 Metabolic relationships of marsupials 9.3 Energetics of locomotion in marsupials 9.4 Responses of marsupials to heat 9.5 Heat loss in exercising kangaroos

10 Aspects of Marsupial Cardiovascular and Respiratory Function 73
10.1 Allometric relationships 10.2 Cardiovascular and respiratory allometry in marsupials

11 Brain and Intelligence 76
11.1 Basic brain structure 11.2 Brain size and intelligence 11.3 Interhemispheric connections in marsupial forebrains 11.4 Learning and problem solving in marsupials

Further Reading and References 84

Index 88

1 Origins of Mammals and the Relationships of Monotremes

The history of mammals is not just concerned with the evolution of the placental mammals and of man himself, but is a complex and paradoxical story involving many other diverse groups. Of these 'other' groups, two still survive and prosper. These are the monotremes and the marsupials. Those with only a casual interest in the origins of mammals often consider the characteristics of these unusual mammals together. However, it should be pointed out that the monotremes and marsupials have little else in common apart from being the only mammalian types, other than the placentals, still inhabiting the earth. The linking of monotremes and marsupials has tended to occur because of the simplistic attitudes about evolution which developed in the latter part of the last century.

The monotremes have been considered to represent the earliest stage of mammal evolution because of reptile-like features in their skeleton and because they lay eggs. Naturally, the marsupials, having pouches and embryo-like young, were seen as the next step in the evolution to the placental mammals, in which marked development of the young occurs in the mother's uterus. The sequential classification of mammals by early workers, such as T.H. Huxley, who called the three groups Prototheria (first mammals), Metatheria (changed mammals), and Eutheria (complete mammals), stemmed from these ideas. Since such a simple picture is now known to be misleading, what are the relationships between the living groups of mammals, the monotremes, marsupials and placentals?

1.1 Origins of mammals

The ancestors of the mammals were reptiles, but these reptiles bore little resemblance to modern reptiles or to extinct forms, such as the dinosaurs, which dominated the Mesozoic Era between 225 and 64 million years before the present (MYBP). Fossils show that the early reptilian ancestors of mammals diverged from the evolutionary lines that gave rise to living reptiles very soon after the origin of the class Reptilia, some 300 MYBP in the Carboniferous period. At this early stage many of the characteristics of modern reptiles had not yet evolved, and so were not present in the reptilian ancestors of mammals. Today's reptiles therefore, cannot be considered as an 'evolutionary stage' preceding mammals.

The history of the animals that gave rise to the mammals, the synapsid reptiles, can be divided into a series of evolutionary radiations, with each successive expansion stemming from a progressive member of the preceding one (Fig. 1–1). Three major radiations were involved in this process: the

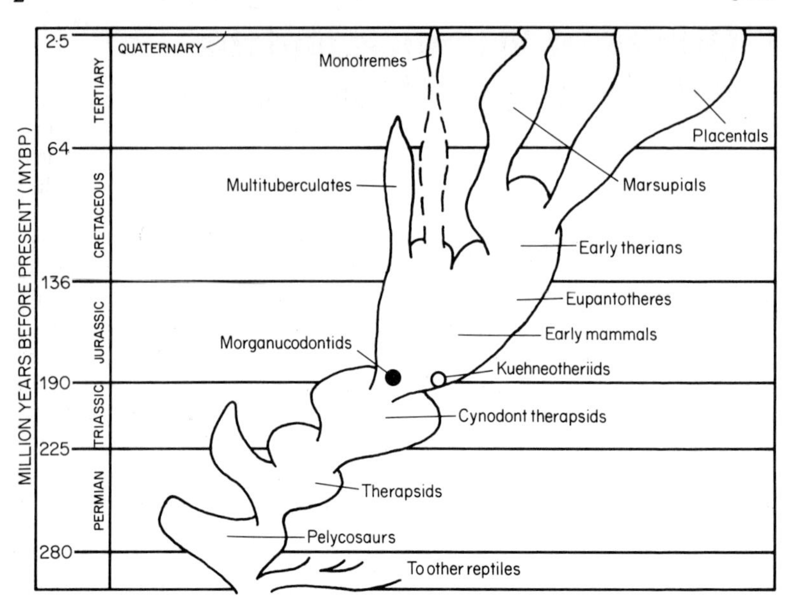

Fig. 1–1 The evolution of mammals from the earliest mammal-like reptiles to modern mammals. The early mammals at the base of the two principal radiations of mammals are highlighted: ● Morganucodontids, O Kuehneotheriids.

Carboniferous and Permian pelycosaurs; the Permian and Triassic noncynodont therapsids; and the cynodont therapsids (or mammalian-like reptiles) of the Triassic. This latter radiation of cynodont therapsids eventually produced the animals we call mammals. The time of this reptile-mammal transition was the late Triassic period about 190 million years ago (Fig. 1–1). Good transitional stages are not known but some of the features now regarded as mammalian probably evolved within these advanced mammal-like reptiles. In the earliest mammals, however, major innovations were introduced that clearly separated them from the cynodonts and opened up for them a large new adaptive zone.

Probably the most important feature of early mammals was the new jaw attachments and modified teeth that had evolved, enabling food to be chewed more effectively. By breaking up the food in the mouth before passing it down the digestive tract it is possible to achieve a faster digestion, and the more rapid delivery of energy to the body. The changes made it possible to accommodate greater metabolic requirements such as those associated with increased activity and the maintenance of a high body temperature. That the trend to increased activity was occurring in the earliest mammals is indicated by the differentiation of the vertebral column to allow its dorsoventral movement. This feature, characteristic of mammals, has not been found in therapsids in which sideways undulations of the vertebral column occurred during locomotion. The development of dorsoventral flexure is a useful adaptation not only for fast running, but also for active climbing.

The changing functional characteristics of the teeth of early mammals are indicated by a change in the arrangement of teeth in the jaw. The late mammal-like reptiles had large canine teeth with smaller premolars and molars behind them, whereas in the earliest mammals relatively large premolars and molars were behind the canines. These mammalian post-canine teeth, especially the molars, had matched shearing surfaces and during the bite food was cut as with scissors. To achieve this precise cutting, one side of the jaw was normally used during one chewing cycle. Chewing and processing food in this way represented a clear departure from the condition in the cynodonts. In these animals the teeth did not shear against each other, rather they bypassed one another so that a gap separated matching upper and lower teeth when the jaws closed. As a result the food of cynodonts could only be torn into larger pieces and digestion would have been slow (Fig. 1–2).

To effectively control the new cutting teeth, complex changes in the jaw and muscles developed. These included the establishment of a new jaw articulation between the lower dentary bone and the temporal bone. This, incidentally, allowed the bones of the old reptilian jaw joint, the quadrate and articular, to be incorporated into the ear to produce a new efficient sound transmitting

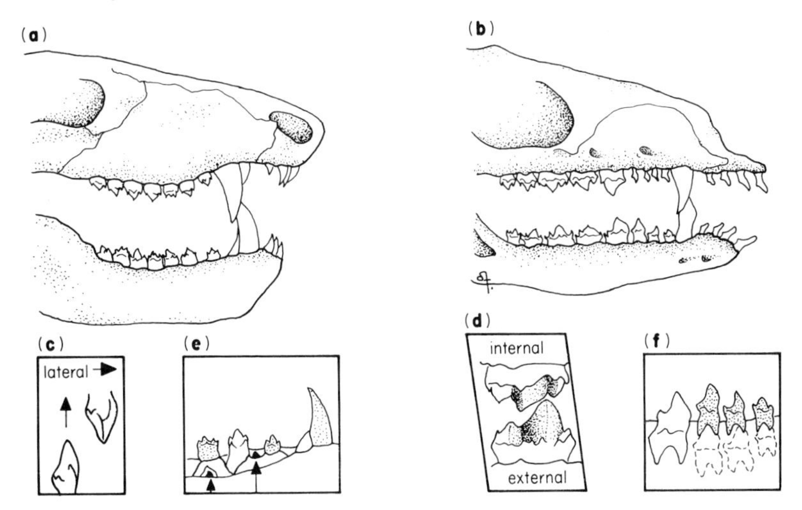

Fig. 1–2 The teeth of (**a**) a late mammal-like reptile, *Thrinaxodon* and (**b**) an early mammal, *Morganucodon*. (**c**) The teeth of *Thrinaxodon* do not cut against each other; (**d**) those of *Morganucodon* do have shearing surfaces as shown by the internal surface of an upper molar and the external surface of a lower molar. Teeth replacement: in *Thrinaxodon* (**e**) new teeth (white) occur between old teeth (stippled) while replacing teeth (arrowed) arise below and near old teeth, while in *Morganucodon* (**f**), the mammalian milk teeth (stippled) are simply replaced, once only, from below. (After Crompton, 1980.)

system. So that the precise shearing relationships of the upper and lower molars would not be disrupted during tooth replacement, a new sequential pattern of tooth replacement was also established. In the mammal-like reptiles new teeth simply erupted between older ones (Fig. 1–2e). In the earliest mammals this pattern changed so that only limited but precise replacement of teeth occurred. Such a change must have been interlocked with the development of another peculiar mammalian characteristic, that of providing the young with milk. Living mammals suckle when they are young and are not reliant on teeth in their early growth stages. Consequently, a large part of skull growth can take place before the eruption of the first teeth, the milk teeth. After weaning, the adult complement of teeth is then produced by a single replacement of the first teeth, together with the addition of several new molars to compensate for some additional growth (Fig. 1–2f).

The earliest known mammals were the first animals to have this characteristic tooth replacement, which points to the initial acquisition of lactation and associated maternal care. The mammal-like reptiles which apparently did not have lactation, required extensive tooth replacement (up to six generations of teeth) to provide the growing young with functional teeth throughout their long growth phase. This was necessary because teeth crowns, once formed cannot enlarge, and in reptiles teeth were required for food processing immediately after hatching. The marked changes in feeding are at the base of mammalian evolution and represent the critical threshold in that evolution. Once they were achieved, the stage was set for the rapid evolution of the many different types of shearing and grinding teeth found among the mammals.

1.2 Relationships of monotremes to other mammals

From among the initial radiation of mammals two distinct groups seem to have arisen which have relationships to modern mammals. One, the Kuehneotheriidae, appears ancestral to the later therian mammals (marsupials and placentals); the other, the Morganucodontidae, is accepted as being ancestral to the monotremes and several extinct groups of other non-therian mammals (Fig. 1–1). Thus monotremes have had an evolutionary history quite distinct from that of the marsupials and placentals for some 180 million years. It is now generally accepted that the monotremes have significant affinities with animals from the first radiation of mammals, that of the 'prototherians'. The marsupials on the other hand are definitely part of the later therian radiation. It is perhaps paradoxical that of the two basic types of early mammals, the Morganucodontidae were the most common and widespread, and yet the details of their relationship to their supposed descendants, the monotremes, are still largely a mystery. The Kuehneotheriidae, however are only known from isolated teeth and jaws but the phylogenetic history of this second group of mammals is reasonably documented. The story of this second major radiation of mammals will have to keep until later, while we consider further those most unusual of mammals, the monotremes.

While the suggested relationship of monotremes to the Morganucodontidae is the most widely accepted, several other relationships have been suggested. These have been reviewed by Mervyn Griffiths (1978) in his recent book on the biology of monotremes. One of these ideas is that monotremes are not mammals at all, but surviving therapsid reptiles. This is an old idea but it has been recently resurrected by paleontologists who would remove most of the early mammals from the class Mammalia and group them as quasi-mammals. At the centre of this argument is really a definition of what is a mammal. Such a definition was attempted by Griffiths, who first listed the attributes of marsupials and placentals, some of which are given in Table 1.

These features of marsupials and placentals occur also in monotremes. It would seem generally inappropriate then to include the monotremes anywhere except with the marsupials and placentals in the class Mammalia unless alternative hypotheses have a strong basis for support. While certain of the

Table 1 Some attributes common to marsupials and placentals, as listed by Griffiths (1978).

(*i*) Articulation of the upper and lower jaw is between the dentary and squamosal bones: reduced teeth replacement.

(*ii*) Middle ear is equipped with three ossicles for the transmission of sound: malleus, incus and stapes.

(*iii*) The young are often small, naked and raised on milk until able to fend for themselves.

(*iv*) Mammary glands with alveoli surrounded by myoepithelium which is responsive to the pituitary hormone oxytocin. Growth and differentiation of mammary glands influenced by ovarian hormones.

(*v*) Temperature regulation assisted by internal heat production, with hair for insulation.

(*vi*) Completely separate right and left sides in heart. The left ventricle pumps aerated blood to the body via the left aortic arch.

(*vii*) Red blood cells are non-nucleated discs.

(*viii*) Blood supplied to the kidneys by a direct renal artery and drained away by a renal vein; a renal portal system is not present.

(*ix*) Nitrogenous wastes are excreted predominantly as urea.

(*x*) Respiration utilizes alveolar lungs and a diaphragm.

(*xi*) Seven cervical vertebrae, the first two modified to form the atlas and axis.

(*xii*) Pelvis has an elongated anteriorly directed iliac blade and a reduced pubis.

(*xiii*) Brains exhibit enormous enlargement of the forebrain, the left and right halves of which are linked by anterior and hippocampal commissures.

above features may be possibly conservative characteristics, and as such also may have occurred in the advanced mammal-like reptiles, the bulk of them are considered to be typically mammalian.

Another idea recently reintroduced and discussed by Griffiths is that initially put forward in 1947 by W.K. Gregory, who suggested that the monotremes have a close affinity with the marsupials. In this scheme monotremes were suggested to be derived from ancient marsupials and to have retained many early marsupial characters. This idea ignores the fact that the therians (marsupials and placentals) have many more characters in common. Similarities between marsupials and monotremes probably only reflect the retention of some conservative characteristics of early mammals.

Evidence based on the structure of the brain case links the monotremes to the early Morganucodontidae but, in addition to this, other information indicates an early separation of the monotreme line from that which gave rise to the therians. Biochemical evidence based on the differences in amino acid sequences of haemoglobin and myoglobin does not support a greater affinity of the monotremes with either marsupials *or* placentals. Estimations of monotreme affinities based on myoglobin suggested a relationship with marsupials whereas calculations using α and β-globin from haemoglobin indicated a closer affinity with the placentals. Such a confusing picture could easily arise if the divergence between the marsupials and placentals occurred well after the monotreme and the therian evolutionary lines separated.

Another idea associated with the scheme of the two basic ancestral groups of mammals, is that the monotremes were closely related to the Multituberculata, one of the most successful groups of early mammals. This proposition was also made on the basis of similarities in the development and organization of the side-walls of the brain. The idea was proposed by R. Brown in 1914 and has been revived recently in the light of new evidence, but there are still some substantial points to be clarified. One factor which makes the search for the evolutionary relationship of monotremes more difficult is their lack of teeth, since teeth have proved to be most useful structures in establishing the evolutionary relationships of other mammals. Only juvenile platypus have teeth and these are shed early in life. They are unique in structure, as are the few teeth from the Miocene (15 MYBP) 'platypus' and give no clues to general relationships.

1.3 Relationships within the monotremes

Modern monotremes comprise two distinct families, one containing the amphibious or semiaquatic platypus (*Ornithorhynchus anatinus*) and the other the terrestrial 'spiny ant eaters' or echidnas, *Tachyglossus aculeatus* and *Zaglossus bruijni*. It has been suggested that a common origin of these two families should not necessarily be taken for granted, although they have many shared unique characteristics. The numerous common features might reflect only shared primitive characters derived from early prototherian ancestors. Such a possibility appears unlikely, however, because analysis of the amino acid sequences of monotreme haemoglobin and myoglobin have shown that the platypus and echidna are more closely related to each other than either is to other living mammals. The separation of the two families is suggested to have occurred in the Eocene or early Oligocene Epoch, i.e. between 52 and 28 MYBP. Recently fossil monotreme material has been found from the Miocene (15 MYBP). This appears to be from a platypus-like animal, now named *Obdurodon insignis* (Woodburne and Tedford). If *O. insignis* really is an ornithorhynchid then this would support a time of separation of the two monotreme families in the earlier Tertiary.

It has been suggested that the two modern monotreme families are survivors of a major adaptive radiation of monotremes, a radiation that occurred in Australia prior to the invasion and adaptive radiation of marsupials. The occurrence of monotremes in Australia only is still a major zoological puzzle. Recent discoveries relating to continental drift and plate tectonics have given potential explanations for the known worldwide distribution of marsupials but have only deepened the monotreme puzzle. If Australia was the centre of an early Tertiary monotreme radiation, and was, as plate tectonics now suggest, connected to other southern continental land masses until Eocene times, why are no monotreme fossils found outside Australia, especially in South America? Is this apparently restricted distribution only a consequence of a

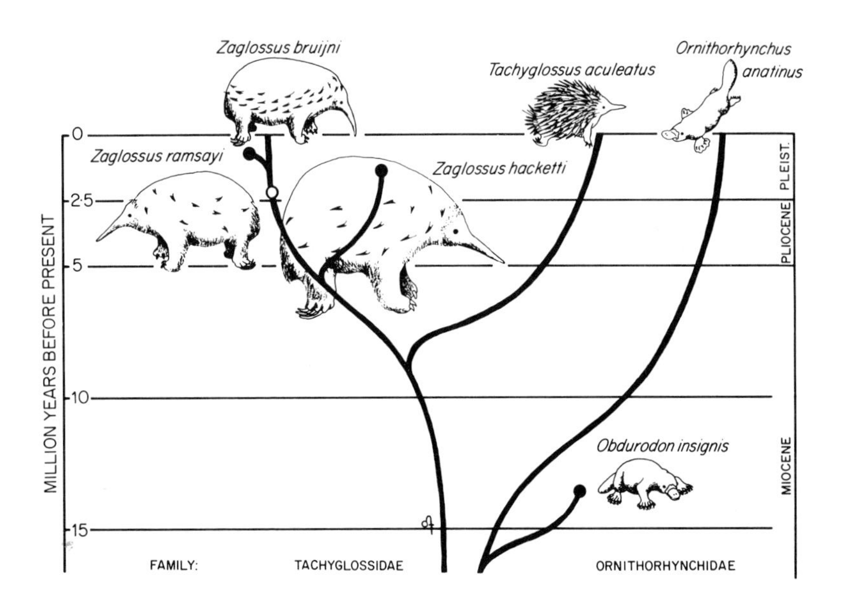

Fig. 1–3 The evolutionary relationships of living and known fossil monotremes. Not figured is *Zaglossus robusta* which is indicated by the open circle, this form being similar to *Z. ramsayi*. (After Archer, 1982, and Murray, 1978.)

poor fossil record or were the monotremes only in Australia and perhaps Antarctica?

While worldwide origins of the monotremes are still a mystery, a picture of their more recent history is now emerging (Fig. 1–3). The platypus line appears restricted and apart from *Obdurodon insignis* other fossil platypus from the Pleistocene are referable to the existing species. The story with the Tachyglossidae is somewhat different. Murray (1978) has shown that there was a small radiation of this family dating from the Pliocene. In the Pleistocene five to six species, which possibly could be divided into three genera, were present in Australia. Two of these still survive, *Zaglossus bruijni* and *Tachyglossus aculeatus*. *T. aculeatus* is the ubiquitous spiny anteater of Australia and New Guinea, while *Z. bruijni* is now restricted to the humid mountain forests of New Guinea and has become specialized as a worm eater. The extinct Pleistocene echidnas included robust and giant forms which have been placed in the genus *Zaglossus* largely because of their size, but Murray has suggested that these forms were probably 'ant eaters' like *T. aculeatus* and unlike the modern worm eating zaglossid. The basis for this evolutionary expansion and subsequent contraction of the Tachyglossidae is obscure.

2 General Features and Natural History of Monotremes

2.1 The platypus, *Ornithorhynchus anatinus*

Much has been written about the discovery of this beautiful little amphibious mammal and the uncovering of its true nature. The first specimens of the platypus which reached England at the end of the 18th century from the penal colony at Port Jackson in New South Wales were regarded with considerable suspicion. Burrell (1927) in his readable account of the platypus aptly illustrates this with an 1823 quote by Robert Knox who wrote: 'It is well known that the specimens of this very extraordinary animal first brought to Europe were considered by many as impositions. They reached England by vessels which had navigated the Indian seas, a circumstance in itself sufficient to rouse the suspicions of the scientific naturalist, aware of the monstrous impostures which the artful Chinese had so frequently practised on European adventurers; in short, the scientific felt inclined to class this rare production of nature with eastern mermaids and other works of art; but these conjectures were immediately dispelled by an appeal to anatomy'.

The name platypus (Latin for flat foot), derives from the first description published by George Shaw in 1799. He originally named the animal *Platypus anatinus* but there have been several name changes, including *Ornithorhynchus paradoxus* (Blumenbach 1800). *Platypus*, having been used for a genus of beetles, was deemed not to have priority, however *anatinus* was acceptable, so the animal become known as *Ornithorhynchus anatinus*. Surprisingly, platypus has survived as the common name instead of duck bill or water mole, names which were used by early Australian colonists.

The extreme specialization of the platypus is reflected in its external features. The streamlined body is flattened top to bottom and covered with very dense brown fur. The snout is shaped like a duck's bill and the tail is flat and broad like that of a beaver. Males weigh up to 2.35 kg and may reach 0.56 m long but females are smaller. The main swimming organs of the platypus are its forelimbs. These are short and stout, with the digits bearing long nails and a web which extends well beyond the nails to form a large fan-shaped paddle (see Fig. 1–3). The hind limbs are also webbed but they are usually kept into the side or act only as rudders, except during rapid swimming. Out of the water the platypus lives in a burrow dug into the bank of its pool. When burrowing, walking or running, the web extensions are folded back and under the long nails, so that the animal moves on its knuckles. The males have no scrotum (testes are internal), but can be distinguished from females by the presence of a large sharp spur on the inside of the ankle of the hind legs; the function of the spur is discussed in section 3.7. In both sexes all urine and faeces are voided through a single cloacal aperture, hence the name monotreme (Greek for one hole).

The platypus feeds in fresh water only and is largely a benthic or bottom feeder. Any food item picked up while the platypus is sifting through the muddy bottom is passed into large cheek-pouches on either side of the mouth, where the food is stored during the dive. When the pouches are full the platypus surfaces, and the contents of the pouches are then ground up between horny pads on the upper and lower jaws. The hard parts of the animals dredged from the bottom, crustaceans, insects and molluscs, are expelled into the water through a series of horny grooves arranged along the edges of the lower jaws. When the platypus dives the eyes, ears, and nostrils are normally shut, so any information about what is to be eaten must come from the touch sensors of the bill. Burrell (1927) suggested the possibility of a 'sixth sense' but this sense appears to be the extremely fine tactile discrimination of the bill.

A study of platypus feeding patterns in the southern highlands of New South Wales indicated a preponderance of the benthic larvae of various insects in the diet over most of the year. In winter however, a significant amount of non-insect invertebrates, crustaceans and worms, also turned up in the diet together with a small amount of vertebrate material. This correlated with a decline in body weight and the fat reserves in the broad tail. Presumably the added energetic stress involved in foraging in cold water necessitates the broadening of the range of types of food eaten. Foraging times are also longer in winter.

Although the platypus is amphibious it is rarely observed on land, its terrestrial activities being associated largely with burrowing. The burrows may be quite long, up to 15 m but are rarely more than 0.5 m below the surface. The entrance is usually above water level and the nest chamber well up the bank above normal flood level. Even major floods do not seem to displace platypus from their normal territory, but how this is managed is not known. Platypus are generally sedentary and maintain themselves in one or two pools of a river although some long distance movements have been noted.

2.2 The echidna or spiny anteater, *Tachyglossus aculeatus*

The echidna is very different from the platypus in its external features. They are related, but each is derived from an old, highly specialized, line. When George Shaw saw the strange spiny animal in London in 1792 he really did not know quite how to classify it and named it *Myrmecophaga aculeata* after the South American anteater because the echidna had been found on an ant hill. He first thought it might form a link between the old world porcupines, which are rodents, and the edentate anteater *Myrmecophaga*. In 1802 a specimen was dissected by Sir Evard Home and its relationship with the platypus was recognized.

Echidnas are stocky animals with powerful short legs, weighing about 3–5 kg. The head is small, about 40 mm wide, but the pointed snout, formed by elongations of the jaw bones, is long, about 75 mm. The upper surface of the echidnas is covered with stout spines (modified hairs) as well as hair, while the flat ventral surface has less hair, and during the breeding season contains the pouch or incubatorium of the female. Like the platypus, the echidna has a cloaca and the male has internal testes. On the forelimbs are large stiff spade-like claws which enable the echidna to dig in forest litter, burrow, and tear open logs and termite mounds to get at its insect prey. The animal is an extremely efficient burrower and when in danger can burrow down rapidly even in hard soil; they are very difficult to dislodge under the circumstances. Perhaps the large marsupial carnivores of earlier times made life difficult for the echidnas. As in the platypus, juvenile females may also have a small spur which is lost in adult life.

The echidna is an ant-eater or termite-eater depending on where it lives. In more arid regions echidnas tend to be termite-eaters but in the wetter areas they concentrate on ants. An interesting example of specialized ant-eating is seen in the attacks on mound building species such as the meat ant *Iridomyrmex detectus*. These attacks, which occur in spring, coincide with the presence of virgin queen ants which comprise almost 50% fat. No matter where echidnas live in Australia or New Guinea they appear to be able to get all their food and water requirements from their specialized diet. The low water requirement is due to the good water conserving capabilities of the kidneys and their nocturnal or crepuscular behaviour, which eliminates the need for water for temperature regulation.

The name *Tachyglossus*, meaning 'fast tongue', is a reference to the long thin tongue of the echidna. The tongue can be extruded up to 180mm at the rate of 100 times per minute and is lubricated with a sticky secretion. Any ants and termites coming into contact with the tongue stick to it and are drawn back into the mouth. The tongue not only catches the food but masticates it as well, since teeth are completely absent. Insects are broken up by the grinding action of tough keratinized spines on the top of the tongue against sets of spines on the roof of the mouth. The tongue is finely controlled during feeding, a notable aspect of which is the stiffening of the tongue. During stiffening, blood flows along a large central artery and out into vascular spaces: arterial pressure prevents back flow and the tongue is thus engorged with blood and becomes stiff. The tongue in this condition is strong enough to break open termite channels in wood. The rapid protrusion of the tongue is brought about by contraction of a system of circular muscles located along the tongue; contraction of these circular muscles squirts the tip of the tongue forwards, all the more so for it being engorged with blood.

2.3 Long beaked echidnas, *Zaglossus bruijni*

Zaglossus bruijni is a large echidna now found mainly in the mountains of New Guinea. It is up to 1m in length; apparently specimens larger than 10 kg are unusual in the wild, but one weighing 16.5 kg was held by Taronga Zoological Park in Sydney. Except for the head and snout its shape and features are similar to those of *Tachyglossus aculeatus*, though its relatively longer legs hold it slightly higher off the ground. The snout of *Z. bruijni* is very long and slender with a downward curve and is an adaptation for probing straight down into the earth from a standing position. This contrasts with *T. aculeatus*, and even some fossil *Zaglossus*, where the comparatively straight and broad snout is used for low angle probing, prying and lifting, in the search for ants and termites nests.

The number of living species of *Zaglossus* was open to debate until recently, but the feature on which the species were separated, the variation in number of claws on the front feet, is now seen to represent a morphological cline. The island of New Guinea runs roughly east-west and animals with five claws are mostly found in the east while four-clawed individuals occur in the west. The spur on the hind foot is also seen in *Z. bruijni*, as in *T. aculeatus*, but there is some doubt whether females lose it on maturity as in *T. aculeatus*. Since other external sexual characteristics are uncertain the retention of the spur by females makes sexing of active individuals difficult.

The food of *Z. bruijni* is primarily earthworms although some ants and termites are eaten. Earthworms are extracted from the soil by the specialized tongue. The tongue is lubricated with sticky saliva and, in general, functions in the same way as in *T. aculeatus* except for marked differences at the tip associated with the capture of different prey. A deep groove extends from the tip of the tongue to about one third of the way back, with rows of teeth or spines in the forward part of the groove. The worm is located by probing with the snout and the tongue is extruded to enclose the worm in the hooked groove. It is then hauled up into the mouth where it is broken up by a similar set of opposable spines as those found in *T. aculeatus*. The long beaked echidna is considered a delicacy by the people of New Guinea and fears are held for its survival as New Guinea develops and its population expands.

3 Reproductive Biology of Monotremes

An attempt to understand the biology of the earliest mammals has been the rationale for many of the studies on monotremes. This approach has proved to be naive because of the failure to recognize the complex nature of evolutionary changes. Like all mammals, the living monotremes consist of a mosaic of characteristics which have responded to varying evolutionary pressures in different ways and to different extents. Many aspects of monotremes are typically mammalian and, if convergence is discounted, this probably reflects the fact that the ancestors of both the monotremes and the therian mammals were 'mammalian' in many essential features not shown by fossils, such as the characteristics of their hair, kidneys, lungs, diaphragm and heart. What then of the supposed primitive features of monotremes? Have there been any non-skeletal characteristics retained in modern monotremes in the ancestral condition, which may yield information about the physiology and behaviour of early mammals? This is a difficult question, but the answer could be yes. The most notable 'primitive' feature is that the monotremes lay eggs, and reptilian-like cleidoic (enclosed) eggs at that.

3.1 Monotreme eggs

The egg laying habits of the platypus and the echidna were a matter of considerable scientific controversy for almost a century after their discovery. Many aspects of their reproductive biology are still unclear. The long confusion associated with monotreme reproduction is entertainingly recounted by Burrell (1927). Conjecture developed around whether the monotremes were:
(*a*) viviparous, that is giving birth to live young which had been nurtured by way of a placenta, as in the case of marsupials and placentals;
(*b*) oviparous, that is the laying of eggs as in the manner of birds and many reptiles: or
(*c*) ovoviviparous, the condition wherein an egg is produced and nutrients for the embryo are largely provided by the egg, but the hatching takes place within the female tract and birth takes place after hatching.

The confusion about egg laying arose because the monotremes are rather secretive animals. Additionally, in the early days they were difficult to keep in captivity and the experts were in Europe, many months voyage from Australia. Many people in Australia, especially the aborigines, were aware of the egg-laying habit of the monotremes, but it was not until 1884 that W.H. Caldwell, a young Cambridge zoologist, made accurate scientific observations of the phenomenon. During a field trip to the Burnett district of Queensland Caldwell obtained an egg from an echidna pouch. He subsequently shot a platypus whose first egg had been laid, but which had a second egg still retained in the uterus. As a result of these and other observations he sent the now famous telegram 'Monotremes oviparous, ovum meroblastic' to a Montreal meeting of the British Association for the Advancement of Science, where it was read on 2nd September, 1884.

The monotremes, then, did lay eggs and these eggs were found to be meroblastic, i.e. they contained nutritive material for a developing embryo. The monotreme egg is therefore equivalent to the cleidoic or enclosed egg of reptiles and birds. For some time after these discoveries, reproduction in the monotremes was considered to follow a reptilian pattern. This is now seen to be only partly true. It appears that reproduction in monotremes differs from the basic patterns in both reptiles and mammals. In 1802 Sir Evard Home described aspects of the morphology of the reproductive system of the platypus, and suggested a close resemblance of some reproductive structures of the platypus to those seen in some ovoviviparous lizards. He suggested that ovoviviparity might occur in monotremes. Although the monotremes were eventually found to be oviparous, there was an element of truth in Home's conclusion.

3.2 Female reproductive anatomy

In both the platypus and the echidnas there are paired ovaries. With the onset of the breeding season the ovaries of both species progressively exhibit an uneveness of surface as the yellow pigmented ovarian follicles containing the eggs protrude up to 4 mm. The ovaries resemble those of reptiles and birds at an equivalent phase. In the platypus however, only the left ovary and oviduct are functional. The right ovary does produce oocytes (primary egg cells), but these do not mature. Unlike the pattern in birds, the non-functional ovary does not undergo compensatory growth in any form after the removal of the left ovary. Both ovaries are functional in the echidna.

The basic arrangement of the female reproductive tract of echidnas is shown in Fig. 3–1. With the exception of the poorly developed right ovary, the tract of the platypus is similar to that of the echidna; the right uterus and fallopian tube of the platypus are only marginally reduced. In the echidna both ovaries are enclosed in well developed infundibular funnels which convey ovulated eggs into long, secretory fallopian tubes. The paired uteri enter separately into the urogenital sinus, there being no vagina in monotremes. The ureters from the kidney also enter separately into the sinus opposite the neck of the bladder. This condition is seen during embryonic development in both marsupials and placentals, but in these animals, as development proceeds, the ureters are progressively incorporated into the neck of the urinary bladder. Exactly how the urine gets into the bladder of monotremes is still not explained. The urogenital sinus of monotremes enters into a cloaca so that eggs, urine and faeces all pass out of the body by the same opening, hence the name Monotremata for the Order.

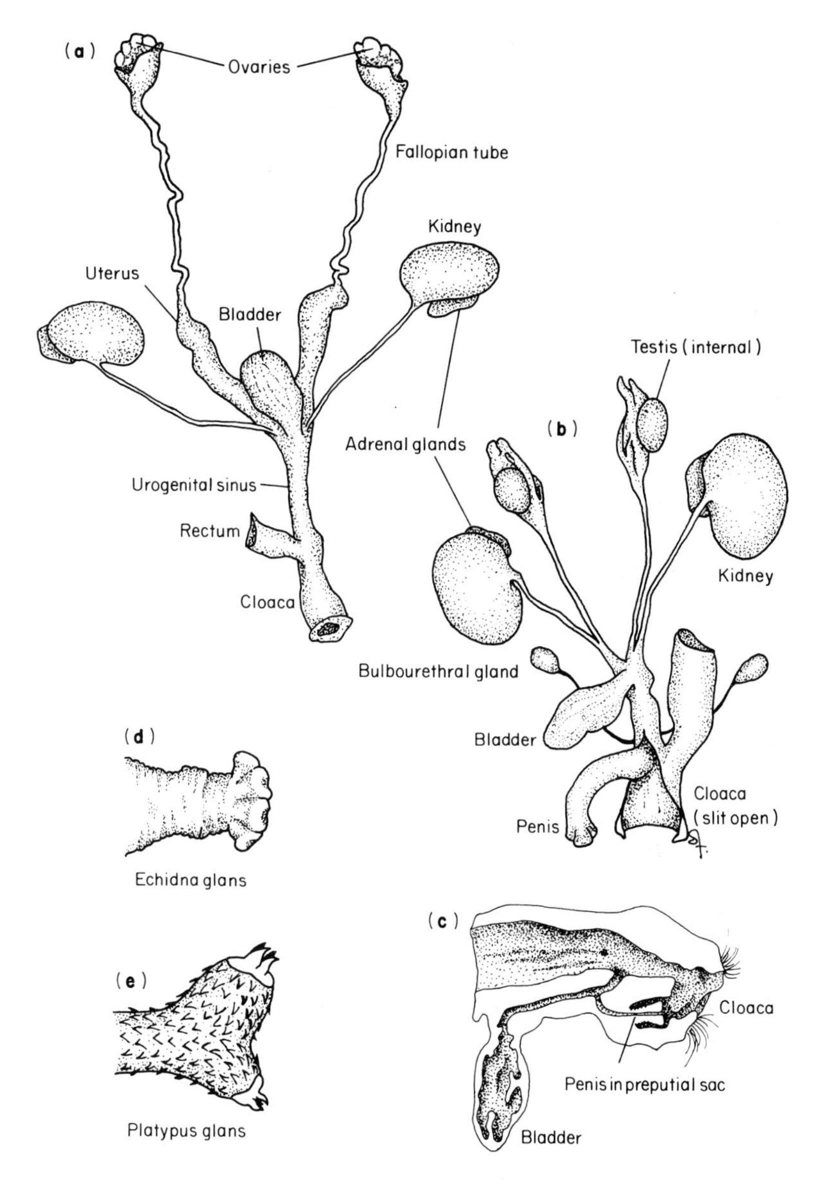

Fig. 3–1 Characteristics of the reproductive system of monotremes. (**a**) Reproductive tract of a female echidna, and (**b**) reproductive system of a male echidna. (**c**) A section showing the resting position of the echidna penis in the preputial sac. (**d**) The glans penis of the echidna, and (**e**) the elaborate glans penis of the platypus. (After Griffiths, 1968, Carrick and Hughes, 1978, and Hughes and Carrick, 1978.)

3.3 Oviparity in monotremes: an insight into viviparity?

The consensus among vertebrate embryologists is that viviparity, as seen in marsupials and placentals, evolved from an oviparous ancestor with a yolky cleidoic egg similar to that of reptiles and birds. What are the features of the reproduction of the monotremes which allow insight into this important development?

During the non-breeding phase the left ovary of the platypus and both the ovaries of the echidnas contain only small follicles, less than 2.0 mm diameter. When the egg is shed it is initially 3–4 mm in diameter. While this is 10 times smaller than a comparable hen egg, it is huge in comparison to the egg of marsupials and placentals, being respectively 15 and 25 times the diameter of the eggs of these other mammals. The egg at this stage has two basic components, the yolky and cytoplasmic contents of the primary egg cell (the vitellus), and two enclosing primary egg membranes. These primary membranes are the cell membrane of the egg itself, and many associated outward projecting microvilli which give the egg a high absorptive capacity. The cytoplasmic component of the vitellus initially occurs as a thin peripheral layer. The nucleus is located within this layer on a small blastodisc, from which the embryo develops. The central yolky globules are synthesized towards the end of follicle growth and are an important source of embryonic nutrients. Neither marsupials nor placentals show a comparable phase of vitellogenesis or yolk production.

Additional egg membranes are deposited during ovarian development. These include the secondary egg membranes, which arise from the investing ovarian follicle cells and include an acellular zona pellucida. Tertiary membranes are the result of the secretions of the fallopian tube or uterus. In this they are typical of the cleidoic eggs of reptiles and birds. After the deposition of these initial shell membranes there is still a marked growth of the uterine cleidoic egg of monotremes, a feature not seen in oviparous reptiles and birds. This growth occurs with the uptake of uterine secretions and is facilitated by the unique structure of the monotreme eggshell which is designed to allow expansion. The thick protective final matrix layer of the shell is only deposited after the egg has reached its full size of approximately 15 mm × 17 mm.

The permeability characteristics of the monotreme egg are shared to some extent with other cleidoic eggs but may be of special significance in understanding the evolution of mammalian viviparity. Permeable shell membranes permitted replacement of the yolky vitellus of cleidoic eggs by nutrients of uterine origin as the major source of embryonic nutrition. The first reduction of the importance of the vitellus is seen in monotremes. In marsupials and placentals the vitellus becomes extremely small, and the embryo is maintained by uterine secretions and nutrients from the maternal blood (Table 2). From an evolutionary point of view the latter system has the advantage that significant maternal resources are not wasted if fertilization fails, or does not produce a viable embryo.

The increasing dependence of the mammalian embryo on nutrients from the

Table 2 Relative importance of sources of nutrients for the embryos of higher vertebrates. (After Hughes, 1977.)

Animal	Ovary	Oviduct	Placenta
Bird			
Gallus domesticus	++++	++	–
Monotreme			
Ornithorhynchus anatinus	++	+++	–
Marsupial			
Trichosurus vulpecula	–	+++	+
Sminthopsis crassicaudata	–	+++	++
Perameles nasuta	–	++	+++
Placental			
Rattus norvegicus	–	++	+++
Homo sapiens	–	+	++++

fallopian tube, then the uterus, and finally the placenta, has occurred in combination with other major changes. These changes include the evolution of a hormone-producing corpus luteum and the insertion of a uterine secretory (luteal) phase into the reproductive cycle. After ovulation, the remainder of the ovarian follicle of monotremes develops into a structure equivalent to the corpus luteum of other mammals. This corpus luteum is believed to be the endocrine gland which regulates the secretory and nutritive activity of the uterus of monotremes. In monotremes, the development of this uterine luteal phase and the substantial intrauterine growth and development of the embryo points the way to mammalian viviparity. Considerable strength is added to this suggestion by the fact that steroid hormones, progesterone and oestradiol-17B, produced by the corpus luteum in other mammals, are present in the platypus at their highest concentrations during pregnancy.

Before claiming a major evolutionary breakthrough for the monotremes and the early mammals however, it should be remembered that most of the basic hormonal patterns involved in this movement to mammalian viviparity were probably already established in the earliest reptiles, unless there has been marked convergent evolution. Specifically, the corpus luteum is present and progesterone production occurs in current reptiles, where they are thought to be involved with behavioural oestrus and ovulation. In viviparous snakes and lizards there is a peak of progesterone production associated with pregnancy. The mammalian patterns then, are not unique but are apparently a development of an ancestoral reptilian innovation.

The production of eggs by monotremes does differ from the reptilian pattern in several significant ways. The marked expansion of the egg in the uterus, associated with the uptake of uterine nutrients, is linked with embryonic growth. A significant part of embryogenesis occurs within the shell while the egg is in the uterus, a period which is thought to be about 28 days, while only 10 days of external incubation occur before hatching. This pattern foreshadows with clarity the viviparity of the marsupials.

3.4 Breeding cycles

Although the egg laying habits of the monotremes indicate their affinity with an ancestral reptilian mode of reproduction, other aspects such as maternal care and lactation are truly mammalian. This is illustrated by the monotreme breeding cycles. The platypus breeds in late winter and spring, from July to October, with animals in northern Australia tending to start breeding earlier than those in the South. The female platypus digs a long nesting burrow into the stream bank; it may be 15 m long but, importantly, is rarely more than 0.5 m below the surface. The entrance to the burrow is usually above water level and the nesting chamber, lined with grass, leaves, reeds and roots, is well up the bank above normal flood level. When the female enters or leaves the nest chamber she plugs the tunnel near the chamber and also at other intervals along the tunnel with earth, presumably to provide protection from predation.

Little is known about egg laying, incubation or hatching of the platypus, but from the work of Burrell (1927) a probable picture appears. The platypus lays 1–3 eggs. Triplets are rare, approximately 7% of the time, while twins are the most common, being found 77% of the time. Single eggs occur about 16% of the time. The eggs are ovoid and when laid are about 16–18 mm long by 14–15 mm wide. The female probably lays the eggs while lying on her back. A sticky mucoid substance covers the eggs so that they adhere together and to the underfur of the abdomen. As distinct from echidnas, the platypus does not have an obvious pouch and the eggs are held between the broad flat tail and the abdominal surface. The muzzle of the curled up mother is also enclosed by the tail and warm humid air is expired into the area, providing the environment for incubation and nurture of the small young.

The incubation period appears to be about 10 days for both the platypus and the echidnas, but definite evidence is yet to become available. To escape the egg the young probably tears the shell by the combined action of the egg tooth and the caruncle, a thickened fleshy outgrowth on the snout. The young platypus are suckled in the nest for perhaps 3–4 months until they are furred and 250–300 mm in length. The time when the young are able to move about and make occasional forays into the water with the mother has not been established, but growth is rapid, and they may reach sufficient size by about 6 weeks.

The echidna young is carried in the pouch for about 55 days, that is until the spines start to grow. Then, understandably, the mother deposits the young in a burrow and returns to suckle it from time to time. The length of suckling is not known precisely but it may be up to 7 months.

3.5 Lactation and mammary glands

Lactation and mammary gland function, like other facets of reproduction in monotremes are still subject to some misconceptions. The existence of mammary glands was doubted for some 30 years after the animals were discovered, and even today it is suggested that they are somewhat primitive. In fact the glands of platypus and echidnas are remarkably similar in their general structure and duct arrangment to those in humans. The areola where the milk ducts emerge however, are covered by fur and no teats are present. The functioning of the glands also follows the general mammalian pattern with the contraction of myoepithelium cells causing milk ejection or 'let-down' in response to suckling. This response is under the control of the normal mammalian milk release hormone, oxytocin.

It has also been suggested that the young do not suck like other mammals but only lick up the milk as it oozes out of a gland onto the abdominal surface. This is quite untrue for the echidna, and presumably also for the platypus, since Griffiths (1968, 1978) has shown that the young echidna actually sucks vigorously, and may take in milk equivalent to about 10% of body weight in a 20–30 minute burst of suckling. The snout of the suckling echidna is flattened on the undersurface at the front end where the mouth is situated. This arrangement is well suited for sucking up milk from the flattened milk patch or areola of the mammary gland. The glands of the platypus and echidna regress in the non-breeding season but are very large in relation to body weight at the height of the breeding season, especially in the platypus. This could explain the fast growth rates of the young platypus.

3.6 Reproduction in male monotremes

The interest in oviparity has tended to obscure the fact that some of the reproductive characteristics of male monotremes are as different from those of other mammals as is oviparity in the females. A general feature of the male genital tracts of both the platypus and echidna is their simplicity compared with that of marsupials and placentals (Fig. 3–1).

The monotremes are testicond, that is, there is no scrotum and the testes lie within the abdominal cavity near the kidneys. The testicular physiology of testicond mammals is little understood, but the widespread occurrence of the scrotum among both marsupial and placentals (perhaps convergently evolved) indicates that the migration of the testes out of the body cavity is of fundamental significance. The explanation generally accepted for testicular descent is the maintenance of a low temperature for sperm production. The scrotum with its special counter-current vascular arrangements can isolate the testes from large changes in deep body temperature, especially those rises that accompany the high heat production of exercise. Most testicond mammals tend to have relatively low body temperatures, and in the monotremes the particularly low body temperatures of about 32°C, may be such as to mitigate against a selective pressure for the evolution of the scrotum.

In male monotremes there is a marked seasonal variation in the activity of the testes, which coincides with the reproductive season of the female. Testes weights begin to increase in early winter and reach a peak in late winter-early spring, when weights may have increased by a factor of fifteen. Regression starts by September and is completed by early summer. This pattern of activity is also followed by the venom gland associated with the unusual ankle spur of the monotremes.

The reproductive tract of the male monotreme is relatively simple (Fig. 3–1). Leading from the abdominal testes are the large epididymides. These are not as intimately associated with the testes as in scrotal mammals and ducta deferentia join the urethra just anterior to the ureters. Only a pair of bulbourethral glands and some small diffuse prostatic glands communicate with the urethra, in comparison to the complex array of glands seen in many marsupials and eutherians. The urethra terminates in the penis, which has a highly elaborate bifid gland. When the penis is not erect it is located within a preputial sack in the floor of the urogenital sinus.

3.7 Poison spur and crural system

A discussion of the poison spur on the hind leg of monotremes may seem out of place here but since this system undergoes a seasonal cycle of growth and regression which is androgen (male hormone) dependent, it is reasonable to assume it has a role in the reproductive patterns of monotremes. The general name for this spur and its venom producing gland is the crural (leg) system. In male platypus the system consists of a kidney-shaped alveolar gland located on top of the upper thigh muscles of each leg. The gland is connected by a duct to the large spur on the heel. The keratinous spur is hollow and its central canal is continuous with that of the duct. Burrell (1927) describes the system and many unfortunate instances of its accidental discovery.

One example Burrell cites is an extract from a letter from Sir John Jamison, dated 1816, to the Linnean Society of London. Jamison wrote: 'I cannot avoid relating to you an extraordinary peculiarity which I have lately discovered in the *Ornithorhynchus paradoxus*. The male of this wonderful animal is provided with spurs on the hind feet or legs, like a cock. The spur is situated over a cyst of venomous fluid, and has a tube or cannula up its centre, through which the animal can, like a serpent, force the poison when it inflicts its wound. I wounded one with small shot; and on my overseer's taking it out of the water, it stuck its spurs into the palm and back of his right hand with such force, and retained them in with such strength, that they could not be withdrawn until it was killed. The hand instantly swelled to a prodigious bulk; and the inflammation having rapidly extended to his shoulder, he was in a few minutes threatened with locked-jaw, and exhibited all the symptoms of a person bitten by a venomous snake. The pain from the first was insupportable, and cold sweats and sickness of the stomach took place so alarmingly, that I found it necessary, besides the external application of oil and vinegar, to administer large quantities of the volatile alkali with opium, which I really think preserved his life. He was obliged to keep his bed for several days and did not recover the perfect use of his hand for nine weeks. This unexpected and extraordinary occurrence induced me to examine the spur of the animal; and on pressing it down on the leg the fluid squirted through the tube: but for what purpose Nature so armed these animals is as yet unknown to me'.

It has been commonly suggested that the spurs are used to hold the female during copulation. However, the presence of the poison, which, while it has not been known to kill a person, has often been fatal to retrieving dogs, would tend to argue against this proposal. The question has been recently re-examined by P. Temple-Smith, and he noted a seasonal variation in the gland that closely paralleled that of the reproductive system. A marked increase in the aggressive use of the spurs during the breeding season was observed, with the aggressive encounters only occurring between males. Temple-Smith also suggested that the spur serves *in lieu* of teeth. The spur may also serve as an antipredator device at the time of the year when the male platypus are active and preoccupied. If the spur was primarily for predator defence, the system might be expected to be active all year and also occur in females.

The spur and crural system probably was a universal ancestral characteristic of monotremes, since a rudimentary spur sheath occurs in juvenile female platypus and also in the echidnas, both in males and females of *Tachyglossus aculeatus* and *Zaglossus bruijni*. The females of *T. aculeatus* tend to lose their spurs with maturity but in the males the gland shows a similar seasonal picture of size variation to that in the platypus. Burrell (1927) considered the spur to be an understandable weapon for male echidnas to use in breeding contests. Because of their impregnable covering of spines, the males could best come together in combat on their hind legs, using the out-turned claws as supports, and with the front of their bodies in contact for their full length. In this position their spurs could be used on the vulnerable underside of their opponent.

The characteristics of the toxin of the crural gland have not been completely determined, although some work on the toxin was carried out by Temple-Smith, who suggested that it is a protein or mucoprotein of high molecular weight. The final determinations of the nature of this unique venom are yet to be carried out.

4 Energy and Temperature Relations of Monotremes

4.1 The characteristics of homeothermy

Mammals, and also birds, are homeothermic animals. Homeothermy allows an animal to maintain a relatively constant body temperature, independent of variation in the environmental temperature. To achieve this it is necessary to accurately balance the body's heat gains and losses. In cold conditions heat losses are initially adjusted by reducing blood flow to the skin and by using fur, feathers or fat for insulation. If these measures are not sufficient to maintain body temperature using only the heat of normal metabolic activity, then extra metabolic heat is produced. This is usually achieved by shivering. In warm conditions and during exercise metabolic activity can produce heat faster than it normally can be lost by convection and radiation. Changes in blood flow are important in increasing heat flow to the body surface, but if this is not sufficient to stop body temperature rising then additional heat loss mechanisms are used. The evaporation of water from body surfaces can remove large quantities of heat and under severe heat stress when environmental temperatures exceed body temperature this is the only significant heat loss route available. Panting, sweating or licking are the mechanisms used for evaporative heat loss.

4.2 Historical attitudes to homeothermy in monotremes

Most hypotheses concerned with the evolution of mammalian temperature regulation have in some way used the thermoregulatory characteristics of monotremes as part of their evidence. Until recently, however, good information about a whole range of monotreme characteristics was not available, and consequently many hypotheses have been poorly based. The aim of this chapter is to give an overall view of what is known about monotremes as homeotherms, without delving deeply at this stage into the ideas about the evolution of homeothermy. This aspect will be considered later, when marsupials are brought into the picture.

Arguments about the thermoregulatory status of monotremes stem from initial studies at the turn of the present century. A. Sutherland in 1887 measured the body temperatures of some monotremes and marsupials and found that they were lower than those of placental mammals. From this he suggested that monotremes and marsupials represented a stage of physiological development intermediate between the homeothermy of the 'higher' mammals and the rudimentary regulation of reptiles. These studies were expanded upon by C.J. Martin in 1902 who, after an examination of the body temperatures, metabolism and thermoregulation of various reptiles, monotremes, marsupials and placentals, tended to support the ideas of

Sutherland. For many years Martin's studies were accepted as indicating that the homeothermic abilities of the different groups of mammals reflected the gradual acquisition of homeothermy as mammals evolved. Inherent in this idea was also an assumption that thermoregulatory and metabolic capabilities evolved concurrently. Results obtained recently differ from those obtained by Martin, but do confirm that monotremes and marsupials have relatively low body temperatures and resting metabolic rates. However, the ability to regulate body temperature over a wide range of environmental temperatures is not necessarily correlated with the level of basal metabolism. As we will see, some marsupials also have a low basal metabolic rate, but are excellent homeotherms. What then is the situation with temperature regulation in monotremes?

4.3 Metabolic relationships of monotremes

A more complex assessment has been made recently of the body temperature and the metabolic characteristics of all three species of monotremes. This shows that all species, when resting in a thermally neutral environment, have similar body temperatures (Table 3). On the other hand, basal or standard metabolism varies considerably between the monotreme species and between the monotremes and other mammals.

At this point it is necessary to comment on why comparisons of metabolic rate are often made at the basal level. The basal metabolic rate (BMR) is the minimal level of metabolism attained in a thermally neutral environment, a postabsorbtive digestive state, and during minimal physical activity. In reality the BMR simply reflects a fundamental level of metabolic organization and activity of animals. Within mammals other levels of metabolism, such as maximal metabolism, are generally related to this basal value. Marsupials, as we will see later (p. 66), vary somewhat from this pattern.

The basal metabolism of the three monotreme species changed with body

Table 3 Average resting body temperature (T_{body}) and weight independent basal metabolism of monotremes compared with those of marsupials and placentals. (After Dawson and Grant, 1980.)

Animal	Average T_{body} (°C)	Basal metabolism (Watts per kg $^{0.75}$)
Zaglossus bruijni	31.7	0.86
Tachyglossus aculeatus	31.3	0.98
Ornithorhynchus anatinus	32.1	2.21
Marsupials	35.5	2.35
Placentals	38.0	3.34

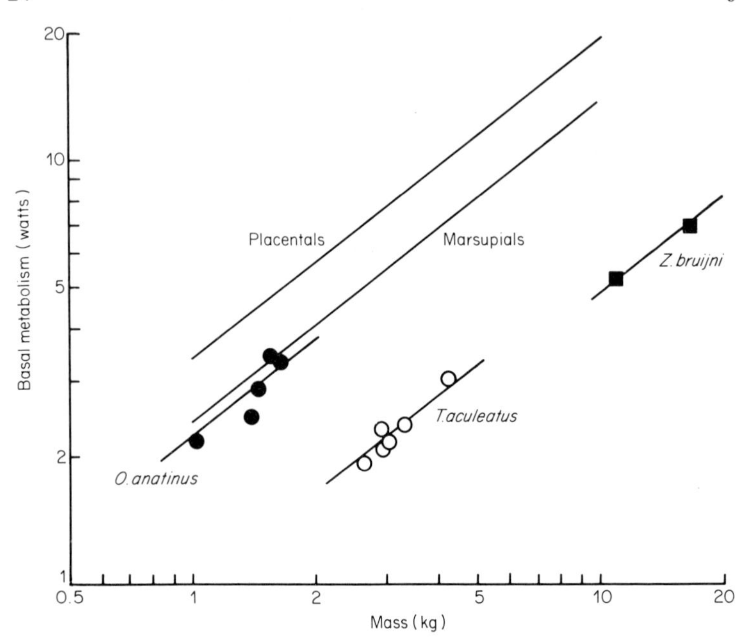

Fig. 4–1 Relationship between basal metabolism and body mass, plotted on a logarithmic scale, for the three monotremes and also marsupials and placentals. (After Dawson and Grant, 1980.)

weight in a manner similar to that seen in both marsupials and placentals (Fig. 4–1). The metabolism of animals does not change directly with weight, but generally at a rate proportional to the 0.75 power of body weight. This similarity throughout the mammals, in the variation of metabolism with weight, is fortuitous because the expression of metabolism per $kg^{0.75}$ allows the characteristic metabolic levels of various groups of animals to be compared directly.

The BMR of *Zaglossus bruijni* and *Tachyglossus aculeatus* is now known to be very low, approximately 25–30% of values normally obtained for placentals. The platypus, however, differs markedly, having a BMR over twice that of the other two monotremes, and only slightly lower than that for marsupials. This difference between the platypus and the echidnas cannot be explained as simple evolutionary divergence.

The two monotreme families diverged as much as 50 million years ago. However, other mammalian groups have been separated as long without such differences occurring between them. Therefore, there may have been marked pressure for adaptive change in the platypus. Many aquatic and semi-aquatic mammals have acquired elevated metabolic rates, apparently to cope with their thermally demanding environment, and perhaps the energetic demands of swimming. This may also be true for the platypus.

4.4 Temperature regulation in the cold

Despite their apparently primitive features, their low body temperatures and low diverse metabolic rates, all the monotremes, as we will see, appear competent homeotherms. Knowledge of their thermoregulatory characteristics does give some insights into the possible characteristics of early mammals.

4.4.1 Temperature relations of the platypus

Within its present distribution in eastern Australia the platypus is found in highland areas where, in winter, it may feed for many hours in water at temperatures approaching 0° – an extreme thermal environment for an animal that weighs only about 1 kg. How does the platypus manage this? If the capability for heat production is linked to basal metabolism, and there are suggestions that it may be, then the heat producing abilities of the platypus may be low relative to those of many placental aquatic and semi-aquatic species. The platypus has overcome this problem however, with adaptive characteristics concerned primarily with minimizing heat loss.

The more important adaptive features of this little monotreme are its cardio-vascular specializations. Most spectacular of these are the complex vascular

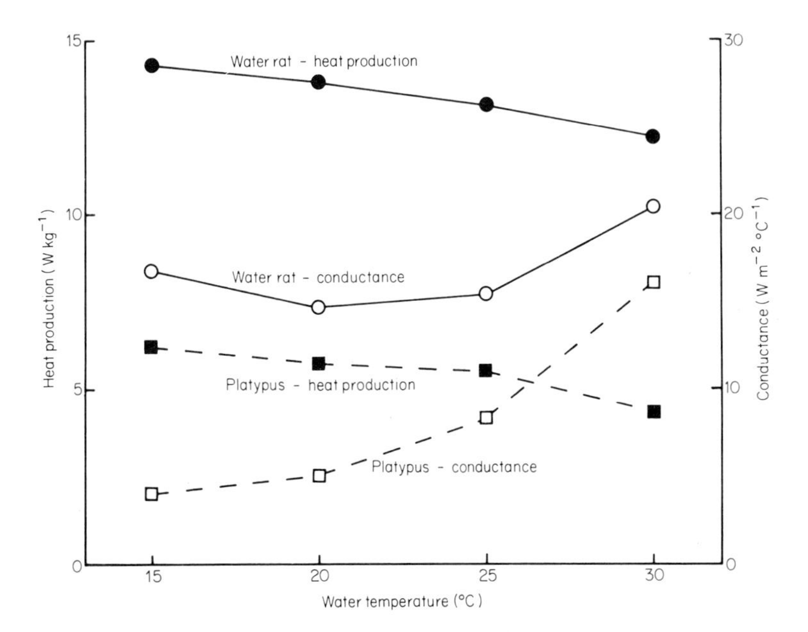

Fig. 4–2 The heat production and characteristics of heat loss, conductance, of the platypus and the Australian water rat in water of varying temperatures. The lower conductance and hence lower heat requirement of the platypus is largely related to its vascular specializations. (After Dawson and Fanning, 1981.)

retia in the blood supply to and from the tail and hind legs. These retia occur where the arteries to the periphery, and veins coming back to the core, divide into many small parallel vessels, with the fine arteries and veins being interspersed. This provides an effective counter-current heat exchange system, the efficiency of which is shown by a comparison of heat loss from the platypus and from the Australian water rat, *Hydromys chrysogaster*, which does not have a well developed counter-current system. When the two species are swimming quietly in water at 15°C, the rate of heat loss from the platypus is only one quarter of that from the water rat (Fig. 4–2).

While cardiovascular specializations play the major role in the maintenance of body temperature in the platypus in cold water, fur also is significant. The fur of the platypus, although it is not particularly deep, is more dense than that of any other aquatic species except that of the sea otter *Enhydra lutris*. In addition to the high density of fibres, the underfur fibres are kinked and the guard hairs are flattened to help to hold a layer of still air next to the body when the platypus is in water. This air layer is essential to insulative integrity in water. Insulation of platypus fur decreased by only 60–70% in water whereas the fur of non-aquatic animals has virtually no insulating ability in water. The overall efficiency with which the platypus reduces heat loss, means that it is able to cope with the aquatic environment with a very low energy expenditure, much lower than the placental water rat (Fig. 4–2).

4.4.2 Temperature relations of the echidnas

Both *Tachyglossus aculeatus* and *Zaglossus bruijni* are also competent homeotherms even though they maintain low body temperatures. They manage this under cold conditions in spite of their very low heat production, which is apparently balanced by a very low rate of heat loss. In fact, *Z. bruijni* from the cold humid mountain forests of New Guinea, has a level of insulation at low air temperatures which is twice that of the platypus under similar conditions. The long dense fur of *Z. bruijni*, and its relatively low surface area to body weight ratio only partially explain this high insulation, and this problem needs further examination.

4.5 Hibernation

Hibernation or torpor has long been thought to occur in monotremes (echidnas). Early workers considered this natural because it was thought that imperfect thermoregulation and a variable body temperature, such as was attributed to monotremes, went hand in hand with hibernation. It has been argued also that the echidnas do not truly hibernate, but enter a form of short term torpor, and then only as a last ditch response to starvation. Some more recent observations however tend to support the occurrence of long torpor or hibernation in echidnas. Griffiths (1968) reported that two animals that had been previously feeding well, became torpid in an unheated animal house in winter. These animals failed to eat for 72 days in one case and 64 days in the other and their body temperatures were less than a degree above that of the environment. At the end of their long fast each echidna fed well and put on weight rapidly. Similarly, torpid echidnas occasionally have been found in the bush. Griffiths observed a young suckling echidna, secreted in a burrow by its mother, for over two weeks, during which period it entered torpor several times. Obviously young echidnas are well prepared to cope with infrequent feedings, after leaving the pouch, by going torpid to save energy.

It has often been stated that platypus do not become torpid, however in 1950 D. Fleay reported that occasionally one of his captive platypus went cold and became torpid, exhibiting a body temperature close to that of its environment.

4.6 Temperature regulation at high temperatures

The response of monotremes to high temperatures is commonly regarded as being poor. Many of the generally held opinions on the thermoregulatory abilities of monotremes are based on the responses of *Tachyglossus aculeatus*. This spiny anteater has an Australia wide distribution, even into the interior deserts. Its ability to cope with such hot environments lies largely in behavioural responses; basically, it avoids the harshest aspects of the thermal environment. In laboratory situations the echidna's response to high temperatures is mainly passive. It does not utilize active evaporative cooling such as panting or sweating, but can increase skin blood flow considerably to help heat loss if a temperature gradient exists from the animal to its environment. The efficiency of this heat loss system is such that a gradient of only 2–3°C from the body to environment can enable dissipation of all the echidna's excess heat production. However, if environmental temperature rises above normal body temperature difficulties arise. Body temperature must then rise to maintain the positive gradient. A body temperature of 38°C can be fatal to the echidna if it has to be endured for long.

These characteristics of *Tachyglossus aculeatus* have been also attributed to the other monotremes but this may not be reasonable. Although *T. aculeatus* has almost no sweat glands this is not the case for *Zaglossus bruijni*, or even for the platypus. Both of these monotremes avoid the heat but also have the ability to sweat. The precise contribution that sweating makes to temperature regulation in these very different species has yet to be determined. For *Z. bruijni* recent studies indicate that sweating may be copious and significant in thermoregulation.

Marked sweating has been observed in the platypus in the laboratory, especially from the poorly furred underside of the tail. It is difficult to see how this animal would normally be faced with a heat problem that necessitated such sweating, considering its way of life. However, much heat may be internally produced by exercise, such as strenuous digging. Knowledge of the neural and hormonal systems which control sweating have given some insight into the evolution of thermoregulatory responses in marsupials and placentals. However, apart from the characteristic role of adrenaline, as seen in most sweat producing systems, nothing is known of the control of sweating in the monotremes.

5 Cardiovascular and Neural Physiology of Monotremes

Some aspects of the biology of monotremes merit discussion not because they are necessarily unique to monotremes but because they highlight specializations on the general mammalian theme. They demonstrate the flexibility of mammals and make us wonder about the possible forms of the animals that may have lived in Australia if there was an early radiation of monotremes.

5.1 Cardiovascular physiology

Heart and circulatory function in the monotremes follows a basic mammalian pattern. They have a four chambered heart, with separate circulation for the lungs, and for the body generally. The lungs and heart lie in the thoracic cavity which is separated from the rest of the visceral mass by the diaphragm. Adaptations of the vascular system of the platypus related to thermoregulation, such as the counter-current rete system in the hind limbs and tail have already been mentioned (p. 25). Other adaptations to their specialized modes of existence are reflected in some of the cardiovascular responses of the platypus and echidna.

Both the platypus and the echidna show a bradycardial response (slowing of the heart) to diving and burrowing, respectively. In these circumstances such adjustments are important factors in the tolerance to reduced oxygen supply. The response involves a reflex slowing of the heart, a drop in its output, and a redistribution of this lowered output of blood. The blood flow through muscle, skin and gut is much reduced. There is a maintained, or increased blood flow to the brain and nervous system, and to the heart itself, so that these tissues get the bulk of the limited oxygen stores. Experiments show that bradycardia develops gradually in the platypus, with heart rate dropping from a pre-submersion level of about 140 beats min^{-1} to near 20 beats min^{-1} after 40 seconds submersion. Recovery at the end of such dives was very rapid and heart rate could jump up to 200 beats min^{-1} within 2–3 seconds. The dives of the platypus are normally relatively short when compared with fully aquatic mammals and seldom are longer than 1–2 minutes.

In echidnas, bradycardia in response to burrowing is much slower in onset, and not as marked in degree, as the diving bradycardia of the platypus. Heart rate during burrowing has been noted to decrease to about half an initial value of some 75 beats min^{-1}. This slow onset probably reflects less severe low oxygen conditions in the soil. The echidna is able to avoid very low oxygen levels (anoxia) by behavioural means. When the carbon dioxide level in the air spaces near the snout reaches 10–12% and oxygen content drops to as low as

7–8%, the echidnas tend to become active and ventilate the soil by raising and breaking the soil above them in a bellows action. In this way the air cavity, which the burrowing echidnas make in the soil around their heads, is kept with a gas content above 10% O_2 and below 10% CO_2. The blood haemoglobin of the echidna has a very high affinity for oxygen, assisting oxygen uptake under low oxygen conditions. The low oxygen requirement (low basal metabolism) of echidnas (see p. 24) is probably also helpful to the animal in these situations.

The low oxygen requirement of echidnas shows up in other aspects of their cardiovascular physiology. Between echidnas and equivalently sized placentals, such as rabbits, the tissue use of oxygen differs by a factor of four. However, the amount of oxygen extracted per volume of blood during each circuit of the body is the same. This results in a much lower total blood flow in the monotreme. The cardiac output of an echidna was 59 ml kg^{-1} min^{-1} against 218 ml kg^{-1} min^{-1} for a rabbit. The oxygen consumption of the platypus is twice that of the echidna and the heart weight of the platypus, relative to body size, is twice that of the echidna heart. Since the heart rates of these mammals do not differ when they are at rest, the platypus heart presumably has the extra muscle to move an additional volume of blood per beat.

5.2 Brain and intelligence

The brain of both groups of monotremes has been studied often with a view to understanding the evolutionary development of the mammalian brain. Such work has been directed to an examination of a 'primitive' mammalian brain, and almost universally the grudging conclusion is that the brain and the intelligence of the monotremes are more developed than would be expected. Pirlot and Nelson (1978) confirmed that monotremes have relatively large brain and neocortical volumes (Fig. 5–1).

The neocortex, the greatly expanded region of the forebrain, is slightly larger in the platypus than in the echidna. This is initially somewhat surprising since the surface of the platypus brain is very smooth while that of the echidna is enormously folded, the latter condition supposedly representing a more advanced state. The platypus' cortical enlargement, however, appears specifically related to its mode of feeding. Most other brain structures are equally developed in both monotremes, except that the echidna has a greater development of the smell areas, the olfactory bulbs and paleocortex, while the platypus has a greater development of the cerebellum, which is concerned with balance. This is as would be expected, smell being important to the terrestrial echidna for finding ants and termites in the ground, but relatively unimportant to the water and burrow-living platypus. The development of the cerebellum and a fine sense of balance would necessarily be important to the platypus in the assessment of swimming and diving position.

The platypus' reliance on its bill for tactile sensory information for underwater navigation and feeding is seen in the enormous development of the area of the cortex associated with the bill sensory inputs (Fig. 5–1b). Most extensive are the inputs from the upper margins of the bill and this fits with the

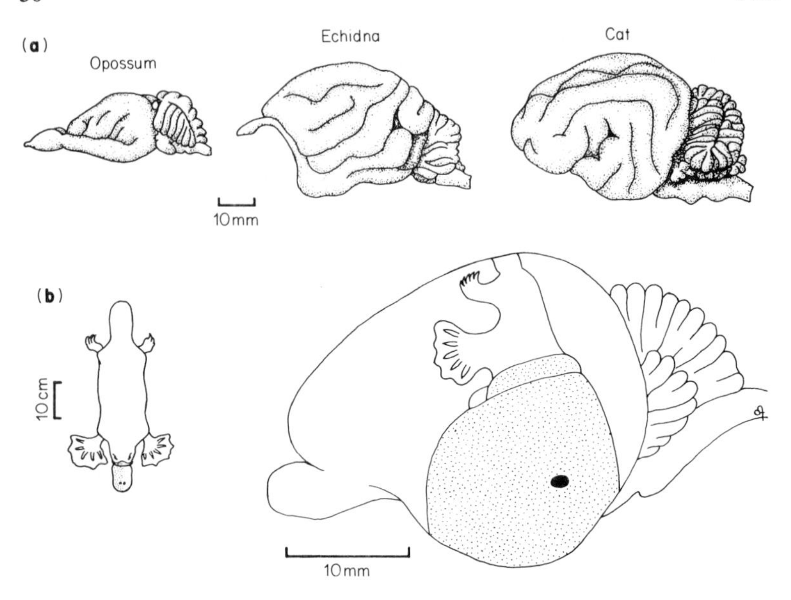

Fig. 5–1 (**a**) Lateral view showing relative size and degree of convolution of the brains of an opossum (*Didelphis marsupialis*), an echidna (*T. aculeatus*) and the domestic cat. The animals were similar in body mass (3–5 kg). (**b**) A lateral view of the brain of a platypus showing the somatosensory areas of the cortex. The figurine highlights the very large somatosensory area (stippled) given over to inputs from the bill relative to actual bill size. (After Allison and Goff, 1972, and Bohringer and Rowe, 1977.)

closure of the eyes and external ear openings during diving. The importance of these sensory inputs from the bill in the platypus is an example of how behavioural specialization is associated with expanded, enlarged cortical activity. For example, the pig has a large part of its cortex dealing with information from its snout and the spider monkey has similar developments for its tail. The smooth cortical regions of the platypus brain are relatively undifferentiated and are generally concerned with specific sensory and motor functions.

The echidna has a relatively smaller proportion of its much folded cortex concerned with sensory and motor specializations and this may reflect a more advanced stage of cortical development than seen in the platypus. In fact the role of much of the large frontal cortex of the echidna is still unresolved and provides neurobiologists with a considerable enigma. Why should an animal like the echidna, with its apparently simple existence, have a frontal cortex comparable with that of man? Echidnas are capable of learning tasks of moderate complexity, at rates equivalent to those of placentals such as cats and rats, however, the enigma of the echidna is further highlighted if the ratio of brain mass to spinal cord-mass is accepted as a useful index of neural organization and intelligence. This idea is based on ratios, which vary from those for fish, in which brain-mass is less than that of the spinal cord, to those for mammals such as that for the cat in which the ratio is 4:1. In primates, the ratio for a monkey is 8:1 and for man, 10:1.

The corresponding ratio for *T. aculeatus* is 6:1. Further studies of learning in echidnas must disclose much important information about their neural capacity and intelligence. In the words of O.L.K. Buckman and J. Rhodes, we must 'modify the quaint, explicitly or tacitly-held views that echidnas are little more than animated pincushions or, at best, glorified reptiles'.

6 Marsupials: Origins and Historical Biogeography

The marsupials are the other of 'other' mammals but the difference between them and the placental mammals is small compared to the considerable difference between marsupials and the monotremes. The marsupials and placentals together represent the latest major radiation of mammals, that of the therian mammals (see Fig. 1–1). If marsupials had a reproductive strategy similar to that of the placentals they probably would not have received special attention, except from academic mammalogists, because in many other respects they are not markedly different from the placentals. Giving birth to embryo-like young and raising them in a pouch or marsupium has drawn considerable attention to the marsupials. The general acceptance of this trait as a primitive characteristic has tended to colour attitudes to other marsupial characteristics, whether or not this was warranted. As with the monotremes, the marsupials will be seen to be a mosaic of characters, some advanced or progressive and others primitive or conservative. It is the aim of this consideration of marsupials to allow them to be seen in their proper relationship to the other mammals.

6.1 The tribosphenic molar

With the coming of the Cretaceous Period (136–64 MYBP) there was great modification of terrestrial environments with the fragmentation and separation of continents. Marked changes were also occurring in the evolution of mammals at this time. The early mammals had molar teeth possessing a series of shearing blades and mechanisms to prevent overclosure and damage to the gums and palate. The prey of these small insectivores was either cut by the shearing blades or torn by the high points (cusps) at the ends of the blades, but the teeth lacked surfaces with which to crush food. In the Early Cretaceous, however, a new type of tooth appeared on the scene; the tribosphenic type of molar. This type of molar appears to have evolved in the earliest therians, the 'therians of metatherian-eutherian grade' which were derived from the eupantotheres (Fig. 1–1). It is from these mammals with their new teeth that the marsupials and placentals were derived. The tribosphenic molar had new shearing surfaces and, notably, a piston-like protocone on the upper molars and a matching talonid basin on the lower molars. The protocone acted to crush food in the talonid basin (Fig. 6–1). The evolution of the tribosphenic molar meant that teeth were available that could both cut and crush food, and these teeth, being molars, were in the posterior jaw region where the strongest muscular forces could be applied.

The development of tribosphenic dentition, and also the initial radiation and

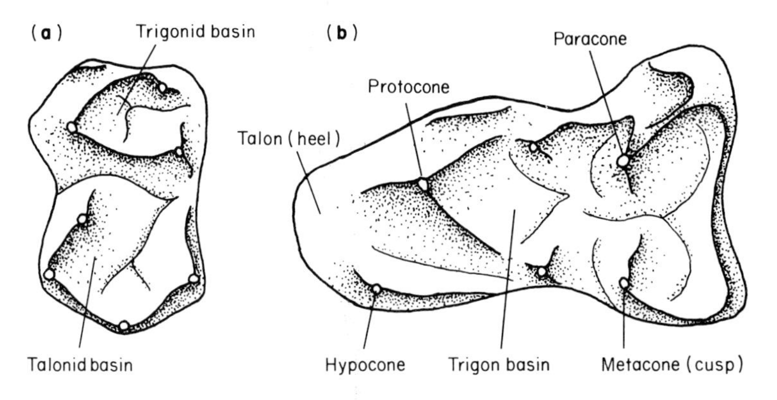

Fig. 6–1 The occlusal surfaces of tribosphenic molar teeth; (**a**) lower and (**b**) upper. The upper molar is reversed so that to visualize correct occlusion it would be folded over the lower molar with the protocone fitting into the talonid basin. Molars generally occlude against a pair of opposing teeth. (After Brown and Kraus, 1979.)

rapid spread of marsupials and placentals seems based on an association with the newly developing flowering plants of the period. The flowering plants, the angiosperms, made their first undoubted appearance in the Early Cretaceous and diversified rapidly to become dominant in the floras of large parts of the world. The link between the evolution of the angiosperms and the origin of marsupials and placentals appears related to changes in the abundance of invertebrate prey. The evolution of many insects and other terrestrial invertebrates was closely correlated with the changes in the flora. The coevolution associated with the development of insect pollination is a notable example.

6.2 Site of origin of marsupials

The fossil evidence concerning the site of origin of marsupials is not as definite as was thought only a few years ago. It was accepted that marsupials evolved in the Early Cretaceous of North America, but now South America is also considered, and other possibilities, notably Australia and Antarctica, have not been completely ruled out. Broadly, the evidence which suggests that marsupials originated somewhere in North America is based on *(i)* the presence of a suitable ancestral stock; *(ii)* the earliest record of morphologically primitive marsupials; and *(iii)* the oldest evolutionary radiation seems to have occurred in the area.

At present the oldest known marsupial fossils come from Alberta, Canada. This Milk River deposit, which dates from approximately 75–77 MYBP, contained at least seven species of marsupials but placental species were rare. Interestingly the multituberculates, the non-therians, were the dominant mammals at that time. In the latter part of the Cretaceous of North America, marsupials were even more diverse than the rodent-like multituberculates, but placentals were still rare. At this time marsupials apparently occupied a variety of ecological niches and ranged from being shrew-sized up to the size of cats. Their teeth, with the tribosphenic molars, indicate that they were primarily carnivorous or omnivorous. Before we assume an important status for the marsupials at this time, it should be remembered that the dinosaurs were by far the dominant animals of this period and the mammals, including the marsupials were 'small fry'. Oceanic barriers, resulting from the breaking up of continents, and the formation of continental seas seem to have restricted the distribution of marsupials to the Americas until the beginning of the Tertiary about 64 MYBP.

Faunistically, the close of the Cretaceous was a time of major changes. Many lines of North American marsupials became extinct, a fate they shared with the dinosaurs. Some didelphid types survived but then only until Early Miocene (20 MYBP). In Eocene times some of these didelphids reached Europe but again survived there only into the Miocene. The cause of the general restriction of the marsupials at the end of the Cretaceous is not understood. Competitive exclusion by the arrival and radiation of new placentals from Asia (which has no marsupials) may be the explanation, but other factors, such as those that caused the contemporaneous extinction of the dinosaurs, may have been involved.

6.3 The Tertiary radiation of marsupials

6.3.1 South America

One of the best documented radiations of marsupials is that seen in the South American record. South America was largely an island continent from the late Cretaceous (65 MYBP) until the late Pliocene (5–7 MYBP) and the fossil record provides a broad outline of what happened during its isolation. During this period, there occurred the evolution and radiation of groups of early placentals together with the spectacular marsupial radiation. That the isolation was not total, is indicated by the fact that, in the Early Oligocene (35 MYBP), two groups of placentals, the rodents and the primates, somehow entered South America. The suggestion that marsupials originated in South America cannot be dismissed, because the appropriate Cretaceous fossil sites necessary to resolve this point have yet to be found. However, by the end of the Cretaceous and the beginning of the Palaeocene, three groups of mammals existed in South America: the marsupials; those very unusual placentals, the edentates; and also the condylarths, the most primitive of the placental ungulates. These early marsupials seem to have been omnivorous, like opossums, but the majority of later South American marsupials were carnivores or insectivores.

The most spectacular forms of this carnivorous radiation were the borhyaenids, which included forms like the placental sabre-tooth tigers of the Pleistocene. These carnivores were in time partly replaced in the biota by large carnivorous ground birds such as *Diatryma* and finally by placental groups which entered the continent from North America. The major marsupial line that continued through the Tertiary of South America was the didelphids, the opossums in their considerable variety.

The marsupial radiations of South America did not produce herbivores of significance except for several lines that became rodent-like. The herbivorous way of life, especially suitable to large animals, seems to have been dominated by the edentates and the descendants of the primitive ungulate condylarths from the Cretaceous. The arrival in the Pliocene (5–7 MYBP) of North American forms, especially new types of ungulates, resulted in a displacement of the old ungulate fauna while the edentates were also much reduced in variety.

The reuniting of the two American continents also resulted in the restriction of the variety of marsupial species in South America, so that today they are represented by less than half of the original families. The extant families are the Didelphidae (opossums), the Caenolestidae (shrew-sized opossums) and the Microbiotheriidae (one species in the genus *Dromiciops*, a small mouse-like opossum). The single microbiotherian is attracting much attention at present because this family has been suggested as ancestral to the Australian marsupials. Even with their restriction in the Pliocene-Pleistocene, the marsupials in South America are still an important feature of the fauna of this large southern continent.

6.3.2 *Australia*

The origins and evolution of the Australian terrestrial fauna have only begun to be vaguely understood in the recent decade. The acceptance of the hypothesis of continental drift has placed geographic limitations on biogeographical hypotheses about marsupial origins and dispersal (Fig. 6–2).

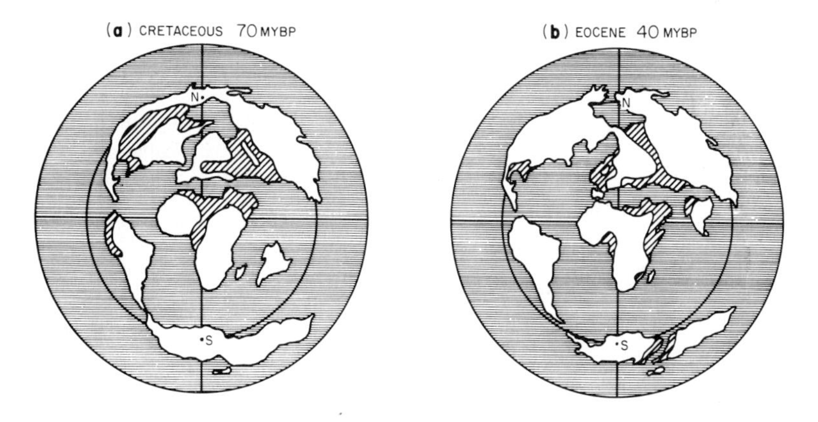

Fig. 6–2 Position of the world land masses in past times. (**a**) Late Cretaceous and (**b**) Late Eocene, as seen in a Lambert equal projection. Horizontal hatching shows oceanic regions and diagonal hatching indicates shallow epicontinental seas. (After Cox, Healey and Moore, 1976.)

Prior to the revival of interest in continental drift in the 1960s, it was held that marsupials had arrived in Australia from North America, their supposed place of origin. They had travelled from North America to northern Asia via a land bridge across what is now the Bering Strait, thence down the coast of eastern Asia. This hypothesis is now untenable, since it is now known that Australia was far to the south of Asia and separated from it by a large ocean gap at the time when this migration would need to have taken place.

The most accepted present hypothesis is a resurrection of an old hypothesis concerning marsupial associations. Early in this century, comparisons of the faunas and floras of Australia and South America led workers to suggest a former connection between South America and Australia, possibly with land bridges via Antarctica. However, these views fell into disrepute because suitable land bridges were not known. At the time the concept of mobile continents was considered too radical and not supported by an acceptable mechanism. The biogeographic story of the origin of Australian marsupials has now been pieced together from plate tectonic models. During the Late Jurassic and Early Cretaceous the great southern continent of Gondwanaland was in the process of breaking up. By the late Cretaceous increasing geographic and then climatic isolation of its component plates led to the development of endemic continental faunas on each. The close proximity of North and South America still allowed some limited exchange of early mammals, however. During this time Australia remained part of a united Australo-antarctic continent, located near the south pole until Eocene times. Dispersal of marsupials into Australia and Antarctica is now considered to have been by sweepstakes dispersal along a volcanic island arc, the Scotia island chain. This island route was formed in Late Cretaceous and Early Tertiary times by a continuous mountain building zone between the Andes to the north and the Antarctic peninsular to the south. This region was at high latitude 60–80°S so there would have been an additional climatic filter effect to sweepstakes dispersal. Such would have been the case even if there had been continuous land connections.

The climate of the Antarctic coasts was cool temperate at these times, being dominated by the southern beech, *Nothofagus*, and old araucarian pine forests. The later development of the Antarctic glacial cap came with the initiation of the peri-Antarctic circulation after Australia's separation and northward drift in the early Tertiary (42 ± 9 MYBP) (Fig. 6–2).

The Scotia island arch and the Antarctic coasts must have acted as a strong filter to the dispersal of mammals, since apparently no placental reached Australia. However, other terrestrial organisms readily found their way across this route and this resulted in a deal of similarity between the flora and some other elements of the fauna of these southern continents. In Australia fossil deposits of an appropriate age and containing mammals are as yet unknown, thus nothing is known of the early adaptive radiation of marsupials that presumably occurred after their arrival. The earliest known deposits containing marsupials are of late Oligocene and Early Miocene age. They show that, by this time, most marsupial groups characteristic of the modern fauna had been already established.

The idea that the Australian radiation had its roots in an invasion from South America by cool temperate-adapted small predaceous or insectivorous forms possibly of didelphine or microbiotherian type appears warranted. An initial radiation in Australo-antarctica however cannot be ruled out as yet. Studies of the similarities of blood serum antibodies indicate that all the living Australian families are phylogenetically closer to each other than they are to any New World marsupial family, suggesting that the Australian marsupials were derived from a restricted common stock. The immigrants apparently entered the region at the beginning of the Tertiary, and later became climatically, and finally geographically, isolated in Eocene times as Australia began its northward drift away from Antarctica.

The northward drift of Australia into warmer latitudes was undoubtedly accompanied by considerable vegetation change especially as the continent entered the dry mid latitudes around 30°S. Widespread beech-pine forests of the early Tertiary became restricted to the southern edge of the continent as temperate forests and warm dry savannahs developed. Later rain forests became established along the eastern ranges of the continent. The development of such climatic and vegetational types would have presented many new adaptive opportunities, especially for herbivorous marsupials such as are now known from the emerging fossil record.

6.4 Marsupial extinctions: Pleistocene and Recent

The extinctions which occurred in the marsupial fauna of South America have been mentioned previously. The event of most note seems to have been the displacement of marsupial carnivores by placental carnivores when South and North America rejoined in the Plio-Pleistocene. In Australia, the older Tertiary fossil record is very sparse. Many species of extinct marsupials are known from only one faunal site and some represent forms that are as yet unique. Phases of restriction and extinction of species will have happened through the Tertiary but details of their actual occurrence have yet to be uncovered. More recent extinction events which the marsupials have suffered are better recorded, however. These occurred during the Pleistocene and in the past two hundred years. Both of these events may have been associated with the coming of man to the Australian continent; in the first instance, aboriginal man, in the latter case, European man.

6.4.1 Pleistocene extinctions

The Pleistocene extinctions in Australia are of general interest because they have been linked with the 'megafaunal' extinctions which occurred elsewhere in the world throughout the late Pleistocene. Such an event happened in North America. There, the arrival and spread of the American Indians was coincident with the extinctions of most species of large mammals, such as the mammoths, giant ground sloths, horses and camels. In Australia, argument has been waged about the role of aboriginal man in the Pleistocene extinctions of large species of marsupials. The marsupials usually discussed in this context are large macropodids and the diprotodontids. The latter were large quadrupedal herbivores, the size of large cattle or rhinoceroses, which survived into the late Pleistocene. Various factors have been suggested as possible causes for these extinctions including extreme climatic change, direct effect of man due to hunting, or indirect effects of man such as habitat change due to fire use, or a combination of these factors.

The suggestion that severe climatic change resulted in the extinction of the largest marsupials is centred on the occurrence of a period of extreme aridity in the late Pleistocene. This period was associated with the last worldwide glacial episode during which much water was locked up in glacial ice. This led to a drop in sea levels and a marked decrease in precipitation.

The glacial maximum was at 18 000 years before present (BP). Between 18 000 and 15 000 BP desert sand dunes extended even to the east coast of Australia in some places. Initially it was thought that extinction of the large marsupials coincided with this period of aridity, however, recent studies have upset this simple explanation because they have shown that much extinction occurred earlier, even before 30 000 BP in a period which was wetter than now.

The pushing of the date of extinctions further back into the Pleistocene was also thought, initially, to have removed them from aboriginal influence. Further work in this field, however, has shown that the aborigines have been in Australia a long time, at least 35 000+ years. This indicates that for a certain time they did coincide with the large marsupials. Unfortunately this time is near the limits of accurate dating by radioactive carbon, C^{14}. The suggestion of aboriginal involvement in the megafaunal extinction is based largely on circumstantial evidence and direct evidence has yet to be provided.

6.4.2 Recent extinctions

The role of man in the extinctions and restrictions of marsupials in the past 200 years is not doubted. This effect, however, generally has not been due to direct hunting, but rather to habitat alteration and deterioration. Most of the species of marsupials which became extinct or much reduced in numbers since the coming of Europeans were inhabitants of the open woodland country. This country is prime farming and ranging country.

The introduction of millions of domestic animals together with land clearing and ploughing has removed many species of marsupials very efficiently. The introduction of additional competitors and predators such as rabbits and foxes has also increased the problems of marsupials in these habitats. The marsupials from desert and mountain forest habitats have been adversely affected by the activities of European man but as might be expected, nowhere to the same extent as those in heavily populated or farmed areas.

7 Marsupial Types, Their Habits and Relationships

7.1 Problems of marsupial classification

There are some 250 species of living marsupials and at least 150 known fossil species, with the latter number being increased rapidly. Uncertainties in the arrangements of marsupials at the higher taxonomic levels of Order and Superfamily reflect deficiencies in the early fossil record, particularly in Australia. At the family level where more information about individual living species is included, classifications have remained relatively unchanged for over a century. Some families group easily into superfamilies and this was the basis of the widely accepted classification of G.G. Simpson (see Table 4, p. 50). Simpson did not group the superfamilies further because of the difficulty in dealing with several significant characters. These difficult characters were 'syndactyly' and 'diprotodonty'.

In syndactyly the second and third toes of the hind foot are closely associated to give a small grooming comb (Fig. 7–1). This occurs in the perameloids (bandicoots) and phalangeroids (phalangers and kangaroos), also in wombats and koalas, while in all other marsupials the five digits are free from each other. Diprotodonty is the reduction of the number of incisors in the lower jaw to a single functional pair, which are enlarged and procumbent, that is forward projecting (Fig. 7–1). This feature is found in the phalangeroids but is also seen in the caenolestoids (shrew-opossums) of South America. Diprotodonty and syndactyly are obvious potential taxonomic characters but the problem is that higher levels of classification based on them do not match. The caenolestoids are diprotodontid but do not have syndactyly, while Australian perameloids are syndactylous but have several pairs of lower incisors (polyprotodonty) like the didelphoids (American opossums) and dasyuroids (Australian native cats and mice).

This incompatibility led for many years to the acceptance of Simpson's superfamily classification, and the grouping of the superfamilies into one diverse Order. The ecological breadth, antiquity and taxonomic diversity of marsupials, has recently made workers realise that the placing of marsupials in one order was inappropriate. In 1964 W.D.L. Ride suggested the recognition of several orders of marsupials, and this scheme has been taken further on the basis of additional evidence, particularly from serological data and chromosome studies. The classification which has gained some acceptance is that proposed by Kirsch in 1977a,b. Some doubts have been thrown on this classification by more recent work, but it still forms a good basis for a discussion of modern marsupial types. In Kirsch's classification there are 16 families of marsupials and since these seem to be natural groups it is pertinent to list these with comments about their member species and various life styles. After this

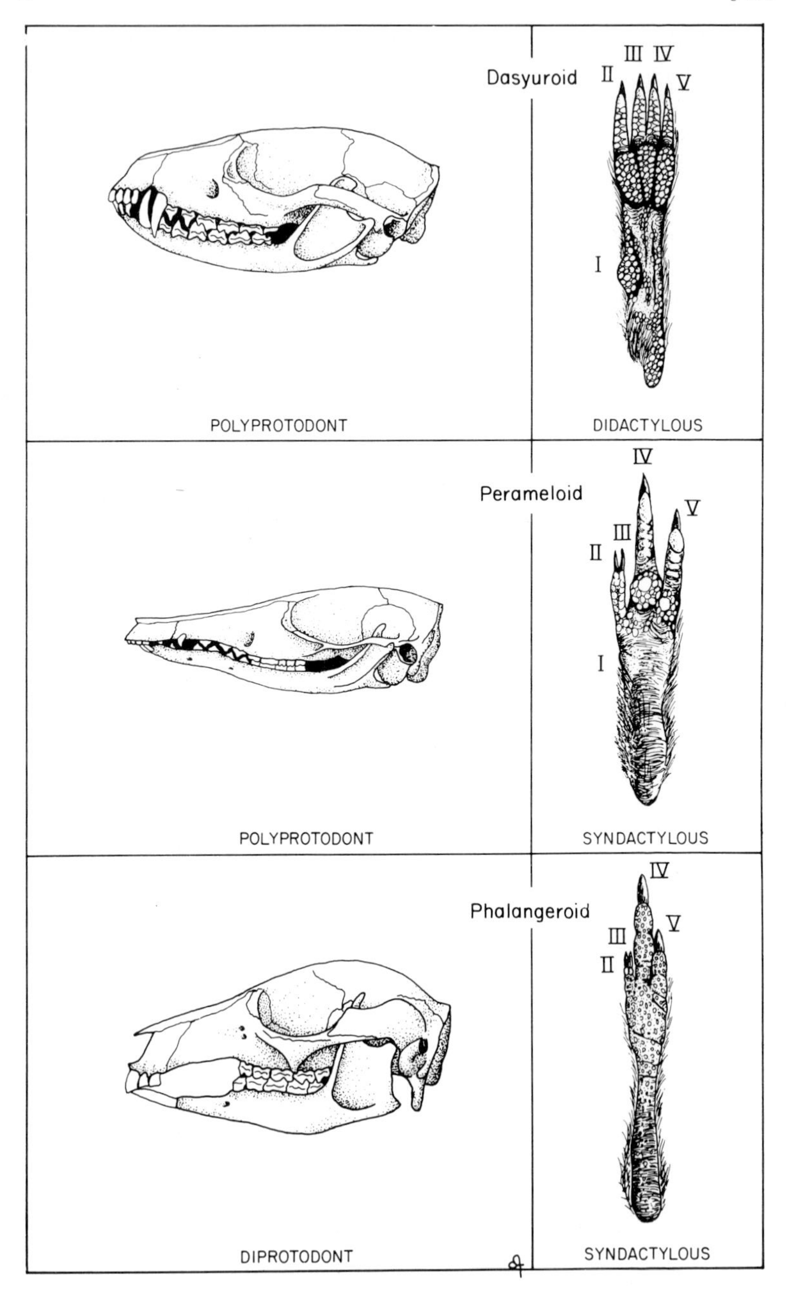

Fig. 7–1 Skull and left hind feet of three marsupial types, showing the distribution of diprotodonty and syndactyly in Australian forms. (After Wood Jones, 1923-5.)

discussion evolutionary relationships and various schemes of classification can be better appreciated. Some of the various marsupial types are shown in Fig. 7–2.

7.2 Marsupial families

ORDER POLYPROTODONTIA

Family Didelphidae (70 species)
Insectivorous and omnivorous opossums of southern and central America and nothern America. In the latter area only one species, the ubiquitous *Didelphis virginiana* (Fig. 7–2q) is present. The past diversity of these new world marsupials was great and not often appreciated. Twelve genera are still extant, the principal ones being *Didelphis* (common opossums), *Philander* (four-eyed opossums), *Metachirus* (four-eyed or masked opossums), *Chironectes* (the yapock or water opossum) and *Marmosa* (the numerous murine or mouse opossums).

Family Microbiotheriidae (1 species)
The single species *Dromiciops australis*, the monito del monte, has become the focus of much interest of late. Several studies have shown it to be very distinct from all other opossums and related to a very early offshoot of the didelphids, the microbiotheriids. Importantly, these animals appear to have a unique arrangement of bones in their feet, a character which they share with the Australian marsupials. Perhaps these small mouse-sized opossums indicate the ancestral forms of the Australian fauna.

Family Dasyuridae (49 species)
The so called native cats, marsupial mice and their allies, such as the Tasmanian devil, make up this Australian and New Guinea family of marsupial carnivores. This group is considered by many to be close to the stem forms from which the Australian radiation was derived. In size they range from the tiny shrew-like species in the genera *Planigale* and *Ningaui*, which may weigh only 4–5 g, to the Tasmanian devil *Sarcophilus harrisii* of 14 kg. Between these two extremes, species may vary from the marsupial 'mice' *Antechinus* and *Sminthopsis* to the weasel-like spotted native-cats of the genus *Dasyurus* (Fig. 7–2k). All are predators, the small ones on insects, the larger ones on a mixture of insects and small vertebrates, such as lizards, with the largest taking some mammals and birds.

Family Thylacinidae (1 species)
The phylogenetic status of the thylacine or 'Tasmanian wolf' (Fig. 7–2e) is contentious and its probable extinction now makes resolution of its position difficult. Its teeth are so like the extinct borhyaenid marsupials of South America (Fig. 7–2a) that this similarity has been frequently cited as evidence of a once closer connection between Australia and South America. Further evidence now indicates that the thylacine is related to the dasyurids, although it has diverged considerably from them.

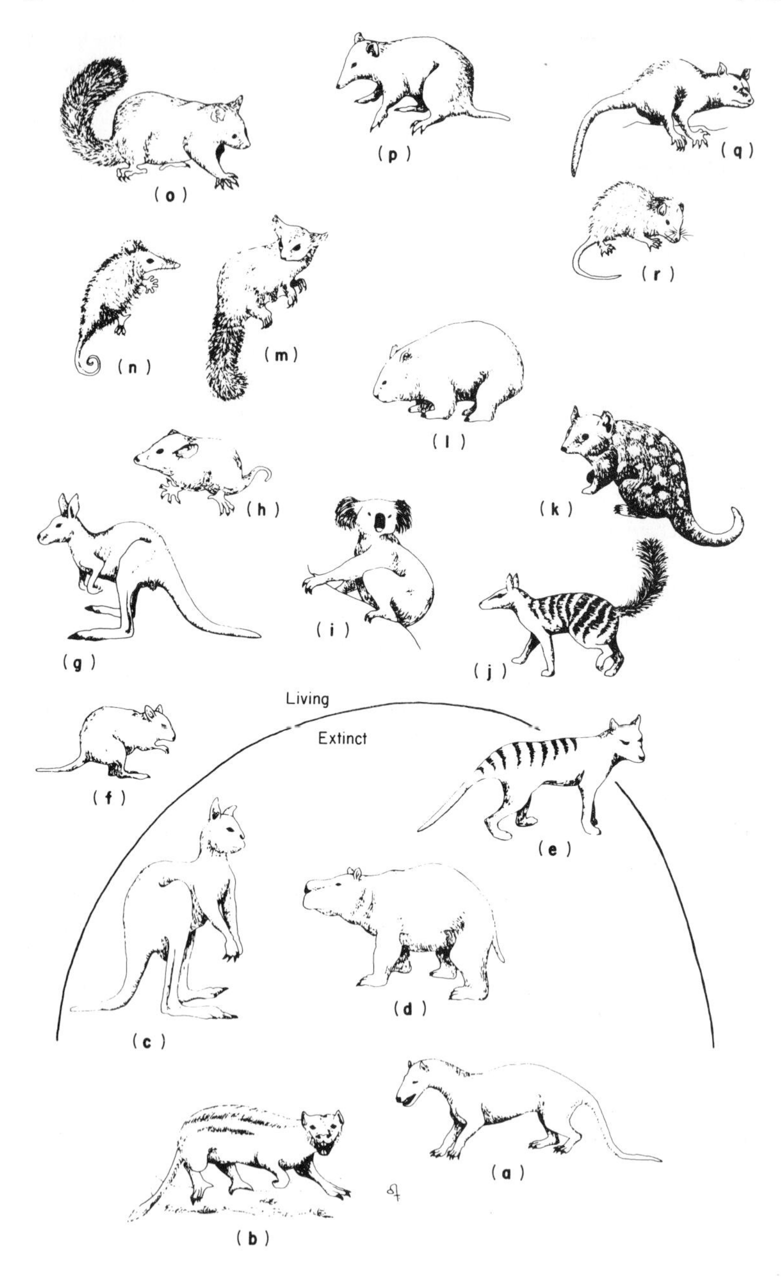
(o)
(p)
(q)
(r)
(n)
(m)
(l)
(h)
(k)
(g)
(i)
(j)
Living
Extinct
(f)
(e)
(c)
(d)
(a)
(b)

Cave deposits suggest that the thylacine disappeared from the Australian mainland with the coming to Australia of the dingo, or wild dog. This apparent ecological replacement occurred about 3000 years ago. The survival of the thylacine in Tasmania reflects this late arrival of the dingo; the separation of Tasmania from the mainland was completed about 12 000 years ago. The reasons for the probable extinction in Tasmania of the thylacine significantly include persecution by the farming community. The 'wolves' of the world seem to have very much suffered from their bad image whenever they came up against farmers.

Family Myrmecobiidae (1 species)

The marsupial anteater or numbat, *Myrmecobius fasciatus*, is another enigmatic animal. It is probably derived from the dasyurids, but is too distinctive to be classified with the 'cats and mice'. The numbat is diurnal and one of the most beautifully coloured of the Australian marsupials with transverse white and nearly black bars across its back, the rest of the fur being a reddish-brown (Fig. 7–2j).

Family Notoryctidae (1 species)

Striking convergence is seen in this pouched marsupial mole, because it resembles the placental moles of other continents. It is small, about 140 mm long, has a horny shield covering the nose, no eyes are visible and there is no external ear beyond a small hole. The marsupial mole is found in the sandridge deserts of central Australia, where it works its way along under the surface of the sand feeding on grubs as well as ants and their larvae. The relationships of this animal remain mysterious, but serological tests and teeth hint at dasyuroid-perameloid affinities.

Family Peramelidae (16 species)

The peramelids are one of the two main evolutionary lines of the bandicoots, the other being the thylacomyids or bilbies. On skull structure, teeth and also serology, bandicoots seem to fit into the Order Polyprotodontia, but their hind feet are syndactylous, a feature characteristic of the diprotodontid marsupials. It is now suggested that the bandicoots may have come from the basic stock which gave rise to the large diprotodontid radiation. In this case a removal of the bandicoots from the Polyprotodontia would appear warranted. The

---

Fig. 7–2 Some representatives of living and extinct families of marsupials. Extinct forms are separated from living species by a solid line; the Tasmanian wolf is not wholly enclosed, denoting controversy about its extinction. (**a**) Borhyaenid, South American carnivore. (**b**) Thylacoleonid, marsupial lion or leopard. (**c**) Macropodid, *Procoptodon* a giant browsing kangaroo. (**d**) Diprotodontid, *Diprotodon* a giant herbivore. (**e**) Thylacinid, Tasmanian wolf. (**f**) Macropodid, small rat-kangaroo. (**g**) Macropodid, large kangaroo. (**h**) Burramyid, pigmy possum. (**i**) Phascolarctid, koala. (**j**) Myrmecobiid, numbat. (**k**) Dasyurid, marsupial cat. (**l**) Vombatid, wombat. (**m**) Petaurid, gliding possum. (**n**) Tarsipedid, tiny honey possum. (**o**) Phalangerid, brush-tailed possum. (**p**) Peramelid, short-nosed bandicoot. (**q**) Didelphid, Virginia opossum. (**r**) Caenolestid, shrew-opossum.

superfamily classification of Simpson (1945) highlighted the unusual taxonomic position of bandicoots.

The Peramelidae are widely distributed in Australia and New Guinea. They are like rather large rats (Fig. 7–2p), hence their common name 'bandicoots', which is not an Australian aboriginal name, but the name of a large Indian rat. They are adapted for a life digging in forest and grassland, scratching for invertebrates and small vertebrates which they locate by a keen sense of smell.

Family Thylacomyidae (2 species)

These bandicoots, the rabbit-eared bandicoots or bilbies, are distinctive and full family status is now accepted. Bilbies are strong diggers and, as distinct from other bandicoots, live in burrows. Behavioural and physiological adaptations suggest a long association with arid and semi-arid environments; food requirements are low and drinking water is not necessary. Strict nocturnality and deep burrows are part of their adaptation to hot arid conditions.

ORDER PAUCITUBERCULATA

Family Caenolestidae (7 species)

The shrew-opossums of South America (Fig. 7–2r) are distinct from other marsupials; the diprotodonty in their teeth is only convergent with that of the Diprotodonta of Australia. A divergence from the didelphid line at least as early as the Palaeocene (64–54 MYBP) is probable. All living caenolestids prefer a densely vegetated habitat such as the wet scrub adjacent to meadows in the high Andes. Like many small marsupials, they live on invertebrates and small vertebrates. All caenolestids lack pouches.

ORDER DIPROTODONTA

Family Phalangeridae (11 species)

Phalangers is the general name for this group of large arboreal diprotodontids, but they are usually referred to as possums or cuscuses. The best known of marsupials to Australian city and town dwellers is the brush tailed possum *Trichosurus vulpecula* (Fig. 7–2o) which has adapted well to cities and suburbs. On the other hand the scaly-tailed possum, *Wyulda squamicaudata*, of the north-west of Australia is one of the least known of Australian marsupials – until 1965 only four specimens were recorded. The cuscuses live in the heavy tropical forests, which extend from Cape York in the northern tip of Australia throughout New Guinea, Celebes, Timor and many Pacific Islands. They have thick woolly fur and long strong prehensile tails. Cuscuses are slow moving and in some respects sloth-like in their behaviour and physiology.

Family Burramyidae (7 species)

The pigmy phalangers or possums are considered to be the most primitive of the diprotodontid marsupials. People are surprised to find possums the size of mice but there is no mistaking them with their prehensile tails and characteristic diprotodont lower incisors (Fig. 7–2h). The principal genus of the family is *Cercartetus*, which has species widespread in Australia and New Guinea. The other Australian pigmy possums are *Burramys parvus*, long known from fossil material, but only recently (1966) found to be still living, and the miniature feather-tail glider *Acrobates pygmaeus*. The pigmy possums are nectar and blossom feeders, but do also take a variety of insects and small invertebrates.

Family Petauridae (22 species)

The ringtails, gliding phalangers and their relatives form a large diverse family of arboreal species. Their evolutionary relationships are still being debated but four or five subgroups are apparent.

The arboreal gliders are striking members of this family, but gliders probably represent 3 convergent lines. As noted earlier, the feathertail glider is related to the pigmy possums, and while the other two types of gliders belong in this family, they have different evolutionary origins. The sugar glider group has affinity with leadbeaters possum while the greater glider is related to the ringtails. The petaurids, or sugar gliders and relatives (Fig. 7–2m), are creatures of the tree tops where they feed on blossoms, tree gums and insects. Gliding is by means of a flying membrane, which is an extension of the body skin. The membrane generally extends to the 'wrist and ankle' and is stretched out to become parachute-like by extending the legs.

The ringtail possums have, of late, been grouped into one genus, *Pseudocheirus*. This group, the members of which have a long, strongly prehensile tail, contains 12 or 13 species. Striped possums also belong to this family and are found in the tropical rainforests of New Guinea and Australia; the one Australian species *Dactylopsila trivirgata* is found in the remnant rainforest of the far north of Queensland. The unusual colour of this black and white striped possum is not its sole claim to attention. In aspects of its morphology and ecology it seems convergent with some of the primitive primates of Madagascar, such as the aye aye. Striped possums have sharp chisel-shaped incisors which are used to tear open the bark of trees and then their much elongated fourth finger is used to extract insects.

Family Macropodidae (56 species)

The diversity of this large family of hopping kangaroo-like animals is the wonder of the radiation of marsupials in Australia and New Guinea. Hopping is rare among higher vertebrates and when it does occur it is usually seen in small desert mammals. No other higher vertebrates, larger than about 10 kg, hop and there is no evidence in the paleontological record other than in Australia of the existence of such large hopping animals.

In the Macropodidae the small rat-kangaroos (Fig. 7–2f) are usually treated together and placed in a separate subfamily, the Potoroinae. The smallest and perhaps most primitive of this group, the musky rat-kangaroo, *Hypsiprymnodon moschatus*, is itself often separated from the other small rat-kangaroos as it retains 5 toes on the hind foot and has a simple stomach and distinctive premolar teeth, but it has many other characteristics in common with the other four genera of rat-kangaroos. It has been suggested that *H. moschatus*, living in its tropical rainforest retreat, reflects the life style of the ancestral macropodids, however it should be noted that its nearest relative is the Pleistocene fossil, *Propleopus* sp., which was the size of a large kangaroo. The other four genera of rat-kangaroos, *Caloprymnus, Bettongia, Aepyprymnus* and *Potorous* have suffered serious restrictions in numbers and distribution since European settlement. Their present day ranges represent only a tiny portion of that of two hundred years ago.

The wallabies as a group are not very different from kangaroos, basically they are only the smaller forms of this diverse kangaroo group. The current groupings of the various wallabies and the kangaroos tend to concentrate on features which show evolutionary relationships rather than just size. This way of viewing the wallabies separates them into about eight basic species groups, some of which are represented by a single distinct species.

(*i*) The larger wallabies, which are closest in appearance to the kangaroos, even the smallest looking like miniature kangaroos. The biggest wallabies are the sandy or agile wallaby from the tropical north and the red-necked wallaby from the forest edges of the south east of the continent. At the other end of the size scale in this group are the tammar and parma wallabies, both of which were once common in their respective habitats but the coming of European man has seen them much reduced in numbers and range.

(*ii*) The swamp wallaby or black wallaby, *Wallabia bicolor*, of the forests of south and eastern Australia was previously thought to belong in the same genus, *Macropus*, as most of the other wallabies. It is now separated because of marked differences in the number of its chromosomes, and differences in its breeding pattern and behaviour.

(*iii*) Pademelons of the genus *Thylogale* are small wallabies about the size of a hare. Their particular habitats are thick scrub and the undergrowth of dense forests, however they may be observed at the edge of the forest in the evening when they come out to feed in clearings.

(*iv*) Rock-wallabies, as their name suggests, are inhabitants of rocky country and cliff faces. The genus, *Petrogale*, is widely spread but occurs in isolated pockets. These animals are incredibly agile and bound among the rock ledges with considerable speed; the soles of their feet have thick pads for cushioning and are roughened to give a firm grip on the rocks. In their rocky habitat the rock-wallabies would be reasonably secure but in some areas feral goats threaten severe competition.

(*v*) Hare-wallabies also have their own special habitat. They are animals of open plains country and were once widely spread in the interior of Australia. Except for the spectacled hare wallaby (*Lagorchestes conspicillatus*) of northern Australia, they are now rare and some species only survive on offshore islands. These animals owe their name to their appearance and to their behaviour, which is similar to that of the European hare in that they have nests in tussocks and bushes and when flushed from these they dart out and make off at speed.

(*vi*) The nail-tailed wallabies, genus *Onychogalea*, are another distinctive type. So called because they possess a small horny nail, rather like a finger nail, hidden in the hair at the end of the tails. The function of the nail is unknown. Three species of nail-tails were once common in open woodland in different parts of Australia, but they are now rare except for the northern species.

(*vii*) The quokka or short-tailed wallaby (*Setonix brachyurus*) holds a special place among the wallabies, because much of the initial scientific work on the biology of marsupials was carried out on this small and unusual wallaby.

(*viii*) The forest wallabies of New Guinea, genus *Dorcopsis*, are notable for the way they hold their tail when standing still. It stands out straight from the body and only the tip curves down to the ground to give support. The reason for this is not known and indeed very little is known about the biology of this group of tropical wallabies.

Kangaroos (Fig. 7–2g) comprise the largest species of the Macropodidae. Three groupings are apparent, the wallaroo-euro complex, the red kangaroos and the two species of the grey kangaroo.

(*i*) Wallaroos are powerful, thickset animals especially adapted for life in mountains or rocky hill areas. Because this type of country is rather discontinuous in Australia the wallaroos have evolved into a number of local types or subspecies. The inland wallaroos or euros have been studied in detail and they seem to be the only large kangaroos which can exist away from some source of drinking water. Under harsh conditions they survive by using a combination of behavioural and physiological responses. The behavioural responses involve the use of cool caves and rock overhangs to avoid the heat load of the desert in summer. They efficiently conserve what little water they get in their food or from chance rainstorms and have the ability to survive on small quantities of high fibre, low protein vegetation.

The Antilopine Wallaroo, *Macropus antilopinus*, is a distinctive large kangaroo from tropical northern Australia and although included in the wallaroo-euro group, it has many behavioural differences. It is a relatively social animal as distinct from the wallaroos, and prefers flatter open woodland country. In many respects in its way of life and appearance it resembles the red kangaroo of the interior of Australia.

(*ii*) Red kangaroos range throughout the open drier parts of Australia. These large kangaroos are gregarious and nomadic and, as might be expected, have many physiological adaptations to accommodate them in their dry and hot environment.

(*iii*) Grey kangaroos comprise two species; the eastern grey, *Macropus giganteus*, and the western grey *M. fuliginosus*. Although Australian bushmen long recognized them as being different, particularly where they overlapped in range, it was not until 10–15 years ago that this was finally accepted by taxonomists. The eastern grey kangaroo occurs in the forests and woodlands of eastern and south eastern Australia including Tasmania, while the western species occurs in a broad band across the south and south west of the continent. Although grey kangaroos are the most common of the kangaroos, little is known about their biology.

Family Phascolarctidae (1 species)
The arboreal koala (Fig. 7–2i) is now placed in its own monotypic family. Its closest relatives are the wombats with which the koala forms a distinct group of Australian diprotodontoids. The koala is one of the most specialized of arboreal mammals. Its diet is wholly tree leaves, mostly of the genus *Eucalyptus* and the koala has a highly specialized digestive system to deal with this unusual diet.

Family Vombatidae (3 species)
While the general relationships of these large burrowing marsupials have been known for a long time, acknowledgement of their affinities with the arboreal koalas is recent. Wombats and koalas are members of an old distinct line of marsupials which dates back tens of millions of years. The wombats and koalas probably separated early in the history of this group and the wombats' closest relatives seem to have been the spectacular giant quadrupedal herbivores which only became extinct a few thousand years ago. The fossil history of the wombats themselves also shows a variety of forms including the enormous species, *Phascolonus gigas*, which had big wide chisel-like upper incisors and probably weighed as much as a cow. Modern wombats fall into two groups; the common wombat of the hilly and mountainous country of the wetter parts of eastern Australia and the hairy nosed wombats which are adapted to the arid plains and grasslands of the interior of Australia. In general appearance wombats are moderately large chunky animals with thick necks and stout powerful legs with strong claws; they may weigh up to 40 kg (Fig. 7–2l).

Family Tarsipedidae (1 species)
The mouse-sized possum (*Tarsipes spencerae*) (Fig. 7–2n) is the most divergent of all the living Australian Diprotodonta. It seems that *Tarsipes* diverged from the other diprotodonts very early in the evolutionary history of this group. The honey possum is a nectar and pollen feeder, which is highly specialized for its association with certain wild flowers, notably the banksias and bottle brushes of the south west of Western Australia. To cope with this food source, the honey possum has a long tube-like mouth and a tongue which has a brushed tip similar to that of many of the nectar feeding birds.

7.3 Evolutionary relationships of marsupials

Courage is needed to give an up-to-date assessment of relationships between living marsupials because it would be soon out of date, given the rate of new fossil discoveries and the development of new taxonomic tools. The various schemes presented here are from a recent synthesis by Michael Archer (1982) and include data from many sources as well as intuitive guesses (Fig. 7–3).

Of the earliest ancestral marsupials from the Cretaceous, the only ones of importance to modern lines of marsupials are the group which gave rise to the didelphoids. The didelphoid radiation in South America in the early Tertiary was spectacular and produced many forms now only recognized as fossils. The carnivorous borhyaenids were notable among such animals. The marsupials

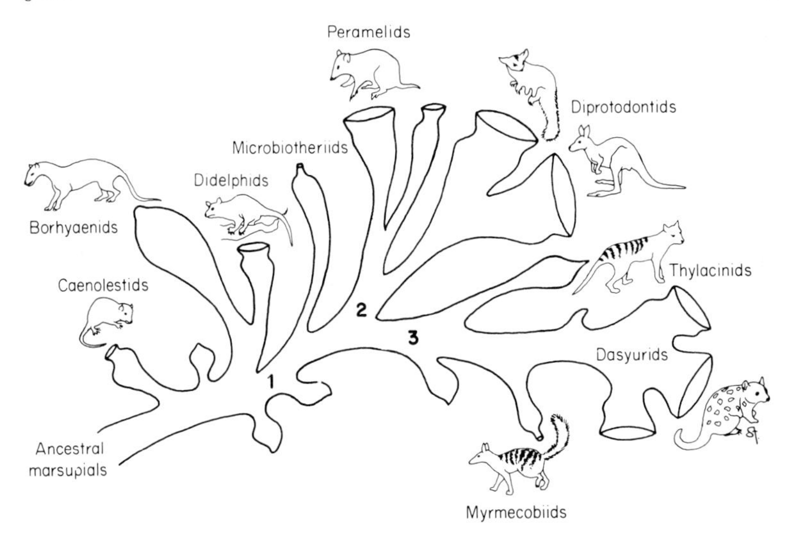

Fig. 7–3 Suggested evolutionary relationships of South American and Australian marsupials. Three points of interest in the evolution of Australian forms are numbered. **1** An early microbiotheriid-Australian group derived from the initial didelphoid radiation. **2** The base of the syndactylous radiation. **3** The base of the dasyurid-thylacinid radiation. (After Archer, 1982.)

now living, which came from this early South American didelphoid radiation are the caenolestids, the modern didelphids and the microbiotheriids. While the microbiotheriids appear to be represented by only one living South American species, they may have been the basal stock for all the Australian marsupials (Fig. 7–3). The microbiotheriids apparently have a unique arrangement of the bones in their feet and they share this character with all the Australian marsupials regardless of their grouping on the basis of other foot characters.

The Australian forms separated into two basic lines on the basis of bones in the hind feet, the dasyurids and relatives which have a simple foot, and the others, all of which have a syndactylous hind foot. This divergence must have occurred in the early Tertiary if the degree of difference between the groups is considered. The basal type for the syndactylous marsupials is thought to be a perameloid-like (bandicoot) animal with the herbivorous diprotodonts evolving later from this type of animal. The variety of forms among dasyurids and related types, such as the thylacine and numbat, suggests that this group has also experienced several major adaptive radiations through time (Fig. 7–3).

Recent changes in attitude concerning the broad classification of marsupials is indicated by the arrangements in Table 4. The first of the classifications is that of Simpson (1945), the second is one proposed by Kirsch (1977). The third arrangement reflects the thoughts at the time of writing by M. Archer, who

Table 4 Some higher level arrangements of living marsupial groups.

Simpson (1945)	Kirsch (1977)	*Current thoughts (1982)
Order	*Superorder*	*Superorder*
Marsupialia	Marsupialia	Marsupialia
Superfamily	*Order*	*Order*
Didelphoidea	Polyprotodontia	Didelphida
Caenolestoidea	*Suborder*	Dasyurida
Dasyuroidea	Didelphimorphia	Paucituberculata
Perameloidea	Dasyuromorphia	Peramelina
Phalangeroidea	Peramelemorphia	Diprotodonta
	Notoryctemorphia	
	Paucituberculata	
	Diprotodonta	

*After M. Archer (personal communication).

suggested to me that this is 'the most rational proposal at this time'. Interestingly, this later suggestion is largely the same as the classification of Simpson, except that the superfamilies have been elevated to the status of orders. The position of the South American microbiotheriids in such a classification is still uncertain.

8 Reproduction in Marsupials

The traditional view is that marsupial reproduction represents a stage of evolutionary development intermediate between the other mammals, the monotremes and the placentals. The gestation period of marsupials is shorter and the newborn young are smaller and in a more embryonic condition than is usual in placental mammals. The tiny young attach to a teat after birth, usually in a pouch, where they are nurtured until they reach a level of development similar to that of young placentals at birth. It is against this background that many features of marsupial reproductive anatomy and physiology were interpreted as being primitive, and limited in flexibility relative to the reproductive characteristics of placentals.

However, the suggestion that there should be differences in the reproductive success of the two therians is only based on zoogeographic distribution patterns and fossil history, not physiology. No recent study has been able to point to the occurrence of significant differences in the relative efficiency of producing young. If a competitive advantage has lain with the placentals it may have operated through non-reproductive features and such a situation may not necessarily exist today. Studies on marsupial reproduction were for a long time undertaken on the assumption that they were inferior. As we will see this has not usually been shown to be the case.

The question often asked is why are marsupials so different in their particular mode of reproduction? Does it represent a stagnation in evolutionary progress, or is it just an alternative and perhaps better reproductive strategy to that used by placentals? The former view has long prevailed but now this is being challenged, principally from an ecological viewpoint, that of the efficient allocation of resources to reproduction. This matter has yet to be resolved, but one thing which is becoming clear is that the initial question is wrongly directed. The question should be: why are the placentals so different? It is they that have diverged from the basic pattern of vertebrate reproduction. A comparative study of marsupial reproduction clearly illustrates this.

8.1 Reproductive anatomy of female marsupials

Among the early reasons suggested for the relative lack of development in marsupial young at birth are simple anatomical limitations due to the structure of the female reproductive tract. The arrangement and development of the embryonic ducts which are eventually concerned with excretion and reproduction differs in marsupials and placentals.

In the embryos of amniote animals (reptiles, birds and mammals) three pairs of ducts are found which are concerned with excretion and reproduction.

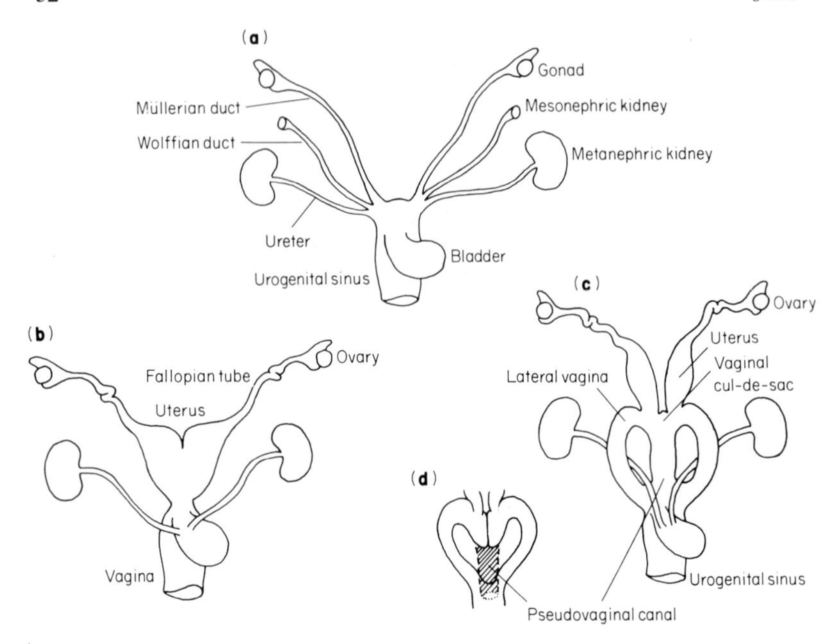

Fig. 8–1 Development of the female reproductive systems in marsupials and placentals: (**a**) sexually undifferentiated stage; (**b**) placental female system; (**c**) marsupial female system, macropodid form with fused vaginal culs-de-sac and open pseudovaginal canal; (**d**) The condition seen in dasyurids with separate vaginal culs-de-sac and a pseudovaginal canal which is only open at parturition. (After Sharman, 1970.)

These are the Wolffian ducts, the ureters, and the Müllerian ducts or oviducts. The Wolffian ducts are initially involved in excretion but eventually become the sperm ducts or vasa deferentia of the adult. The Müllerian ducts provide the fallopian tubes, uterus and vagina. During the course of development in marsupials, the ureters run to the bladder between the Müllerian ducts, whereas in placentals the ureters enter the bladder by passing lateral to or outside the Müllerian ducts (Fig. 8–1). Since the ureters also take the path between the Müllerian ducts in reptiles and monotremes, the marsupials retain the ancestral condition.

It has been suggested that the placentals have had a selective change in embryogenesis which resulted in the alternative path for the ureters. This change allowed fusion of the Müllerian duct derivatives to form a large uterus and vagina and thus permit the production of large advanced young. Such a simple explanation for the difference between placentals and marsupials is less than satisfying. Fusion of the Müllerian ducts to give a large uterus is not necessary to achieve large young. Some placentals, even some ungulates which give birth to advanced young, have completely separate uterine horns and marsupials also show some fusion and development of the reproductive tract.

In marsupials there are two lateral vaginae up which the sperm travels on insemination. In ancestral mammalian forms birth or egg laying presumably occurred through these lateral vaginae, the homologues of the midline vagina of placentals. In modern marsupials birth occurs through a midline passage, the pseudovaginal canal, the short cut to outside from the culs-de-sac formed where each lateral vagina loops around a ureter at the base of the uteri (Fig. 8–1c). The most primitive vaginal condition in living marsupials is to be found among the smaller carnivorous marsupials. In the small dasyurids and some didelphids there is still a septum, separating right and left vaginal culs-de-sac, present throughout life (Fig. 8–1d). In these forms the pseudovaginal canal is transitory, being formed anew at each birth.

Among the kangaroos and wallabies however, the epithelia of the large vaginal cul-de-sac and urogenital sinus become continuous at the time of the first birth. A permanent midline birth canal is thus formed, which is analogous to the midline vagina of placentals. In view of this condition it is difficult to accept that simple anatomical constraints would have limited the size of marsupial young if there had been strong adaptive pressures for the birth of large young.

8.2 Reproductive cycles of marsupials

A feature of the gestation period of marsupials is its short duration. Why does the developing young marsupial spend such little time, in some cases only 12–13 days, in the reproductive tract? What is achieved in such a short time?

In female mammals the reproductive tract has two primary functions. The first is the reception and transport of spermatozoa to the egg so that fertilization can occur. The second function is to produce egg coats and shells and provide nourishment for the developing embryo. The complex sequence of events which make this possible is called the oestrus cycle. There are two phases of the cycle; the pro-oestrus or follicular phase and the luteal or secretory phase. These names refer to the stages which occur in the activity of the ovary during the cycle, but they are reflected in the whole reproductive tract.

The broad sequence of changes in the ovaries and genital tract are similar in most marsupials (Fig. 8–2). The breeding cycle starts with the pro-oestrus phase during which the ovaries enlarge and follicles grow and mature. Cell division and growth of the uterine epithelium and secretory glands occur at this time. The vaginal complex also increases in size and secretory activity to enable the reception and transport of seminal fluid. These initial changes are controlled by oestrogen levels, since they can be induced by the injection of oestrogen in intact females or in those with the ovaries removed. Oestrogen induced changes reach their peak at about the time of oestrus or 'heat', when copulation may occur. Ovulation occurs 1–2 days after oestrus. Growth of the uterine glands continues at this time and for a further short period. The epithelium of the vaginal complex regresses, its job of transporting and storing spermatozoa being finished.

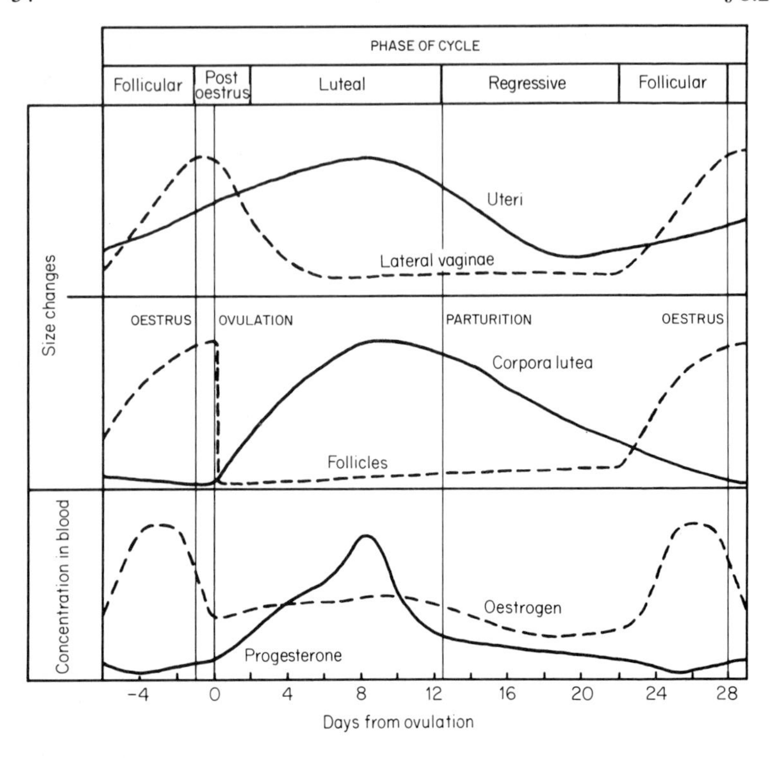

Fig. 8–2 Broad sequence of changes in the ovaries and reproductive tract during the oestrus cycle of the virginia opossum. (After Hartman, 1923, and Harder and Fleming, 1981.)

The brief post-oestrus proliferative phase in the uterus is followed by a secretory (luteal) phase which is under the control of the corpus luteum. The corpus luteum forms from the wall of the ruptured ovarian follicle and the endocrine luteal cells produce the hormone progesterone which is responsible for the onset of the secretory or luteal phase of the cycle. Just as the proliferative phase can be induced in castrated females by the injection of oestrogen, so the luteal phase with its secretory activity can be induced by the injection of progesterone. The luteal phase, which varies in length in different marsupials, is followed by a regressive phase during which there is a reduction in size, complexity and secretory activity of the uterine glands. Anoestrus follows in monoestrus marsupials, i.e. those which have only one oestrus per breeding season. In polyoestrus marsupials however, the regressive or post-luteal phase grades into the next proliferative phase (Fig. 8–2).

Insemination and a resulting pregnancy do not interrupt the oestrus cycle of marsupials. This is a significant factor of marsupial reproduction. Pregnancy does not inhibit the continuation of the oestrus cycle but lactation does, the ovarian inhibition being mediated via the suckling stimulus. If lactation is prevented by the removal of the young soon after birth, then the next oestrus re-occurs when expected. In most marsupials, gestation does not extend beyond the luteal phase, but the situation is very variable. In others, particularly the kangaroos and wallabies, gestation may extend well into the next pro-oestrus phase, with birth recurring only a day or two before the next oestrus and mating. Rarely, gestation extends beyond the next oestrus; in the swamp wallaby the oestrus cycle is 32 days while gestation takes 35 days. A strict alternation in ovulation of the right and left ovaries, however, allows this small break in the marsupial pattern. In macropodids the extension of gestation to occupy most of the oestrus cycle is thought to be due to some foetal or placental stimulation of the uterus subsequent to the decline of the effect of the corpus luteum. The stimulation is probably hormonal in nature but whether it is the embryo or the placenta which is involved is not clear. However, pregnancy still fails to inhibit the oestrus cycle and this indicates that the extra-ovarian hormone secretion during pregancy is not comparable in significance to the situation in the placental mammals. Within the amniotes the prolonged maintenance of pregnancy is unique to the placentals and again it is apparent that marsupials have largely retained the basic ancestral pattern.

8.3 Foetal development and placental relationships

8.3.1 Eggs and egg membranes

In the adaptation to life on land, the evolution of the cleidoic egg was a major development. It freed amniote vertebrates, from the reptiles onward, from the need to return to water to lay eggs. It also provided sufficient resources for the young to be more or less independent at hatching. The basic feature of the developing cleidoic egg is its series of functional compartments, the amnion, yolk sac and allantois (Figs 8–3, 8–4). The amnion grows over the embryo and encloses it in a cushioning fluid-filled sac. The yolk sac is an outgrowth of the embryonic gut and encloses the yolk while another posterior outgrowth of the embryo gut, the allantois, functions to store metabolic waste products. Both yolk sac and allantois become partly vascularized, especially where they come in contact with the shell. The shell is differentially permeable and a substantial exchange of oxygen, carbon dioxide and water vapour takes place between the embryo's blood and the atmosphere in these vascularized regions. In most birds and reptiles the allantois grows far more rapidly than the yolk sac and becomes the main respiratory organ before hatching.

The egg membranes and shell membranes of marsupial eggs follow the basic amniote vertebrate pattern and in essentials are similar to the conditions described for monotremes. Their structure and persistence are important to discussions concerning the fundamental differences between marsupial and placental reproduction. Primary egg membranes are derived from the egg itself and include its cell membrane together with its outwardly projecting microvilli. Secondary membranes arise in the ovary from the follicle cells and include the acellular zona matrix. Tertiary egg membranes are secretions of the oviduct. The inner tertiary egg membrane consists of a mucoid coat (in birds this is supplemented by protein components to give the egg 'white'). The outer tertiary membrane is a shell membrane of structural protein, keratinous in nature. In mammals it is the tertiary egg membranes which have been much reduced. The fully formed shell membrane of monotremes has an overall thickness of about 70 μm. In marsupials the shell membrane rarely exceeds 10 μm and the mucoid layer is also greatly reduced. Importantly, the marsupial shell membrane persists for much of the pregnancy, only rupturing in the last third of gestation. Placentals have eliminated the shell membrane, and the mucoid coat is only seen in rabbits and their relatives.

Another obvious feature of mammalian eggs is the reduction of the yolk, reflected in the size of the egg. Monotreme eggs on ovulation are about 4.5 mm diameter while the eggs of marsupials are only 0.12 to 0.28 mm in diameter; the eggs of placentals tend to be smaller still, 0.07 to 0.15 mm in diameter, although they overlap the marsupial range in size.

8.3.2 Embryonic development

Fertilization occurs in marsupials before the outer keratinous shell membrane is formed around the egg. In the fallopian tube many spermatozoa are entrapped by the mucoid coat but only one penetrates to the egg nucleus. During the initial cleavage of the egg in marsupials, the dividing cells (protoderm) become arranged into a single layered sphere and their yolk is extruded into the centre of the sphere (Fig. 8–3). This is the unilaminar blastocyst and it may contain from approximately 50 to 1000 cells. In non-mammalian amniotes the protoderm is arranged as a disc on the relatively flat surface of a large yolk. Cleavage in placentals follows another different pattern, with the initial formation of a solid ball of cells, the morula, the blastocyst cavity, the blastocoele being formed later (Fig. 8–3).

Further development occurs when the unilaminar blastocyst begins to differentiate into the two layered or bilaminar blastocyst. In the region of the blastocyst which is to become the embryonal area, some of the protoderm cells thicken. On division, some of these thickened cells migrate inward to form a continuous lining of flattened cells, the endoderm. In the bilaminar blastocyst the outer protoderm layer becomes the ectoderm; the endoderm forms the inner wall of the yolk sac.

Following the formation of the bilaminar blastocyst the third germinal layer, the mesoderm, develops (Fig. 8–3). The mesoderm makes its appearance below a slight thickening in the embryonal ectoderm, the primitive streak. The mesodermal cells divide and spread outward from the primitive streak between the ectoderm and the endoderm until they reach about half way around the wall of the blastocyst. In this way the upper half, or embryonal side of the yolk sac becomes three layered and the lower half remains two layered. As the blood vessels develop only in mesoderm, the upper part becomes the vascular yolk sac and the lower part the avascular yolk sac. The blood vessel at the limit of the vascular yolk sac is the sinus terminalis.

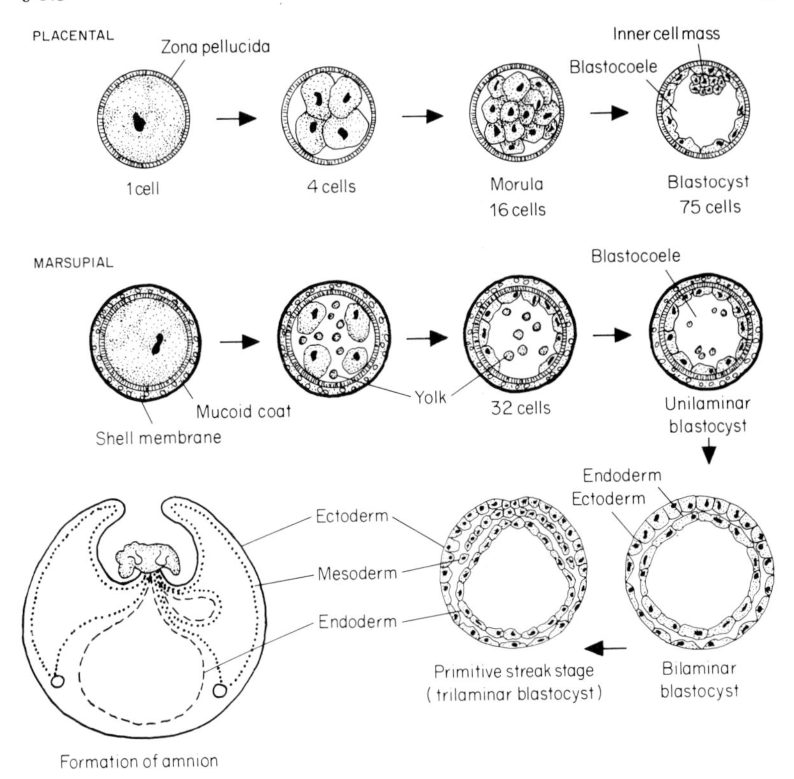

Fig. 8–3 Early development of the fertilized eggs of marsupials and placentals. The formation of the amnion is also shown. See text for details of the various stages. (After Lyne and Hollis, 1977.)

The amnion is formed when the embryo sinks beneath the surface of the blastocyst dragging ectoderm with it. At this stage the mesoderm is separated into two layers, an outer nonvascular (somatic) and an inner vascular (splanchnic) layer. The somatic mesoderm and the ectoderm fold over the embryo and fuse to produce the amnion (Fig. 8–3), which is then lined with ectoderm and covered by somatic mesoderm. The formation of the amnion also results in the formation of another compartment within the blastocyst, the extra-embryonic coelom (Fig. 8–4b). The outer wall of the extra-embryonic coelom is known as the chorion and is comprised of the inner layer of somatic mesoderm and the outer ectodermal layer. The ectoderm remains continuous around the blastocyst and is called the trophoblast. It has been suggested that the trophoblast in placental mammals has special characteristics which allow the marked extension of gestation in these mammals.

A special feature of marsupial gestation is the relatively slow growth and development of the embryo in the period prior to its intimate contact with the uterine epithelium. During this time, the free vesicle phase, the outer shell membrane is still intact. Once the shell membrane breaks down, development and organogenesis are rapid. Although gestation varies in length among marsupials, this final growth phase is similar in length in most species regardless of size, being between 4 and 8 days.

8.3.3 Foetal membranes of marsupials

Associated with the reduction in yolk content of mammalian eggs, as indicated by their reduced size, there has been an increased dependence on the oviducts and uterus for nutrients. In addition, development inside the mother necessitated the development of some mechanism for gas exchange, particularly if growth was to be rapid. In the cleidoic egg of reptiles and birds these functions are carried out by the outgrowths of the embryonic gut, the yolk sac and the allantois. While the yolk sac has a primarily nutritive role and the allantois functions to store metabolic waste products, both may be vascularized and develop a respiratory function when they contact the shell. In the mammals the yolk sac and allantois have also been variously adapted to provide the embryo's needs.

When foetal membranes such as the embryonic yolk sac or allantois are applied to the uterine wall, the structure formed is referred to as a placenta. The pattern seen in the monotremes is considered to represent the simplest solution, and appears to be an adaptation of the reptilian pattern. Thus, in monotremes the allantois, covered with vascular mesoderm, retains its respiratory function both before and after the egg is laid and the yolk sac still provides the nutrients. While the embryo is developing in the uterus additional nutrients to those provided by the egg yolk are supplied via the vascular yolk sac which is applied to the outer shell membranes. This structure is called a yolk sac placenta, and through it transfer of nutrients is facilitated. A similar yolk sac placenta occurs in many marsupials, particularly before attachment while the shell membranes are still intact. However, when the marsupial embryo attaches to the uterine wall, even loosely, a variety of further developments occur in the placental arrangements of many species (Fig. 8–4).

Placental mammals do not use the yolk sac for a nutritive function, but have a placenta derived from the allantois to provide all their exchange needs (Fig. 8–4d). This complex chorio-allantoic placenta forms within the first days of gestation. The first stages of its formation involve growth of the allantois (endoderm covered with splanchnic mesoderm) outward from the hind-gut into the extra-embryonic coelom, as occurs in other amniote vertebrates. However, in placental mammals (and some marsupials), there is further growth and the allantois comes into contact with the chorion (somatic mesoderm and trophoblast) and fuses with it, generally over a limited area. Where the fusion occurs the trophoblast invades the wall of the uterus, the epithelium and connective tissue of which breaks down (Fig. 8–4c and d). In this way the foetal blood supply in the chorio-allantoic placenta comes into close contact with a rich maternal blood supply, thus enabling a rapid exchange of nutrients, gases, excretory products and other vital substances between foetus and mother.

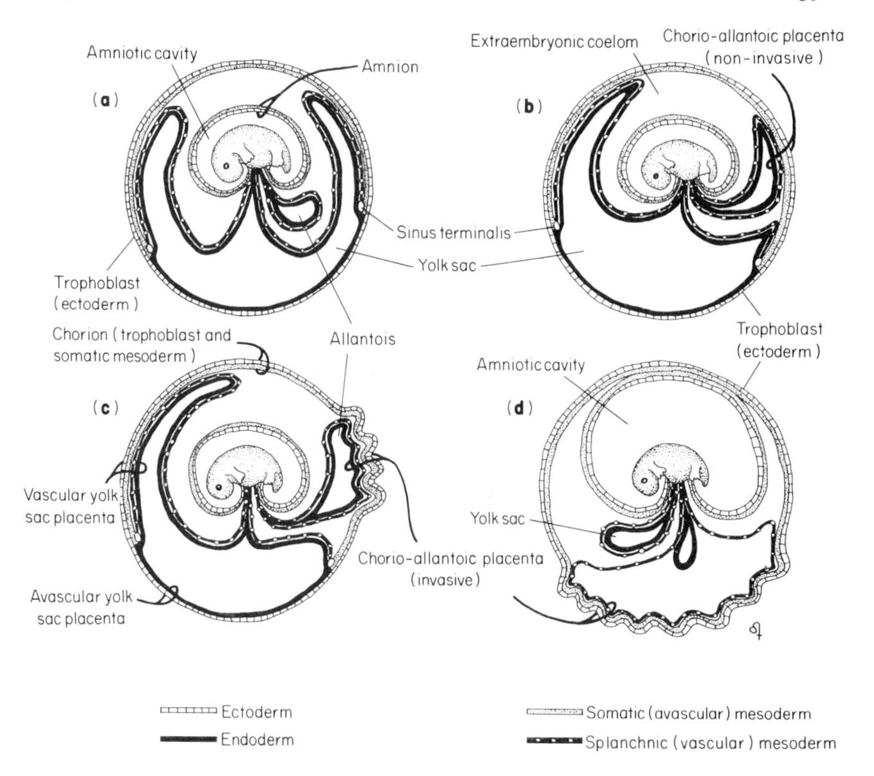

Fig. 8–4 Relationships of the foetal membranes seen in some marsupials and placentals. (**a**) Simple yolk sac placenta – *Didelphis* and many other marsupials; (**b**) non-invasive chorio-allantoic placenta with yolk sac placenta – koala; (**c**) invasive chorio-allantoic placenta with yolk sac placenta – bandicoots; (**d**) invasive chorio-allantoic placenta only – placentals such as humans.

Within the marsupials several patterns of placental development occur. The basic and perhaps ancestral marsupial condition is a yolk sac placenta only, with generally only a late and tenuous attachment of the embryo to the uterine wall. Although the allantois may expand rapidly near full term in this condition, it does not fuse with the chorion, the outermost foetal layer, because it is prevented from doing so by the expanded vascularized yolk sac (Fig. 8–4a). Such a non-invasive yolk sac placenta occurs in many marsupial groups, including didelphids, phalangerids and macropodids. However, in some species, the trophoblast of the yolk sac placenta shows some degree of implantation, that is, penetration into the uterine endometrium. Again examples are widespread throughout the marsupials. Both an absorptive and synthetic function has been suggested for the yolk sac placenta.

Another type of placentation found in marsupials is a combination of a yolk sac placenta together with a chorio-allantoic placenta. This type of placentation is best developed in those odd diprotodontids, the koala and the wombats. In this case (Fig. 8–4b), a true 'chorio-allantoic' placenta is present but there is no implantation, and no actual erosion of maternal tissues takes place. Such an arrangement probably facilitates some additional exchange across the placenta but the level of this exchange is not known.

The bandicoots occupy a paradoxical position in the phylogeny of marsupials, having syndactylous hind feet but polyprotodont teeth. They also have a reduced yolk sac placenta and a well developed chorio-allantoic placenta (Fig. 8–4c). In *Perameles nasuta*, the long nosed bandicoot, there is actual implantation at the final stages of gestation when growth is very rapid. The implantation takes place in a similar manner to that in many placental mammals. Trophoblastic villi of the chorion invade the uterus and erode it and the end result is a close approximation of the foetal and maternal capillaries. A significant difference between the chorio-allantoic placentae of bandicoots and those of placental mammals occurs late in gestation. The trophoblastic cells of bandicoots fuse with maternal multinucleate cells and thus disappear as a functional layer, allowing even closer contact between the foetal and maternal blood. The trophoblastic layer remains intact to the end of gestation in placental mammals (Fig. 8–4d).

It has been suggested that placental mammals have developed the chorio-allantoic placenta to produce large advanced young. While this may be its function, it is informative to look at the use to which bandicoots have put their chorio-allantoic placenta. Bandicoots apparently used the greater exchange potential of this type of placenta to markedly reduce the time it takes to produce young. Their basic reproductive pattern is similar to that in other marsupials, but faster. The gestation period of 12–13 days is very short but the young are relatively well advanced. Their pattern of growth is also speeded up after birth; the young have a short nursing period and become independent at approximately 50 days, compared with 250 days for some kangaroos.

If it is not an inability to provide a 'true' placenta that results in the use by marsupials of the pouch to provide nurture during development, what other reasons are there? One suggestion involves the properties of the embryonic trophoblast. The trophoblast has been proposed as being involved in the crucial adaptation leading to prolonged gestation in placental mammals in that it may provide the barrier that keeps the pregnant animal from mounting an immune reaction to reject the 'foreign' tissue of the embryo. The developing embryo is foreign to the mother in the sense that it consists of tissues derived genetically from the father as well as the mother. It has been suggested that the marsupial trophoblast has not evolved this immuno-protective ability and the embryo is only protected from potential rejection by its inert eggshell membranes, which are of purely maternal origin. This idea has some support because the shell membranes are retained until late in pregnancy in marsupials. This retention of the shell membranes is then suggested to place limits on the size of the marsupial young because the final phase of gestation must be short, so that the young is born before a fully fledged immunological attack can be launched on it.

Hypotheses which suggest different characteristics for the trophoblast in placentals as compared with marsupials are generally based on the difference between placentals and marsupials in the first phase of the cleavage of the egg. The placentals diverge from the usual amniote pattern by the formation of the solid morula before the formation of the hollow blastocyst (see Fig. 8–3). Lillegraven (1975) has argued that the placental trophoblast (ectoderm) derived from the morula is not homologous with that of marsupials and has fundamentally different immunological characteristics. The evidence for this is a matter of debate. These immunological based hypotheses are weakened however by the finding that repeated inseminations of female wallabies by the same male does not seem to affect their overall reproductive efficiency. If there was no immunological barrier or suppression in marsupials during the attachment phase of gestation, then sensitization to the male derived antigens would be expected to increase with each mating. That this does not happen tends to undermine the simplicity of the 'trophoblast hypothesis'.

8.4 Relative efficiency of marsupial reproduction

While marsupial reproduction is often considered as an example of evolutionary stagnation, with minor adaptations to circumvent some of their problems, there is another point of view. This involves the relative efficiencies of the marsupial and placental systems, and the sexual control of the breeding. Recently several authors, notably Pamela Parker (1977), have argued that marsupials are not necessarily inferior in their reproductive pattern and that the marsupial mother may have an advantage in comparison with the placental. This advantage is suggested to occur because the marsupial mother retains a large degree of control of her own reproductive mechanism, rather than surrendering a considerable degree of control to an invasive organism (i.e. the foetus) which contains only half her genes.

The marsupials put fewer resources into their young during gestation than do placentals. Also during the long suckling period they may be able to terminate their reproductive attempts more easily if the chances of their offspring's survival become low. Marsupials then may be able to adjust reproduction and population growth rates more easily than placentals. If this was so then varying environmental conditions could be exploited by marsupials with lower overall rates of investment of energy and other resources.

The flexibility of the marsupial system in this regard is best demonstrated by the phenomenon known as embryonic diapause, which is a special feature of the reproduction of many species of kangaroos. During diapause, a viable embryo is carried in the uterus for long periods with its development arrested at the blastocyst stage, when it consists of from 70 to 100 cells and is only about 0.25 mm in diameter. The advantage of this adaptation can be appreciated when seen in the context of the kangaroo's overall reproductive pattern. Fertilization and pregnancy, as noted before, do not alter the course of the oestrus cycle in kangaroos, and another oestrus and mating can occur again soon after birth.

The fertilized ovum from such a post birth mating develops to the blastocyst stage, but then becomes dormant if the newly born young reaches the pouch and begins to suckle. Lactation inhibits both blastocyst development and the oestrus cycle. In a species such as the red kangaroo if the newborn young in the pouch survives normally, the blastocyst will remain dormant for about 200 days. Thereafter, development of the diapausing embryo resumes and within 30 days birth takes place; the previous inhabitant of the pouch, now grown large, is evicted shortly before the new offspring enters its shelter. Birth again may be followed by oestrus and mating, with another blastocyst resulting. Since the young (called a joey) ejected from the pouch may continue to suckle for another four months, the female kangaroo may have three offspring in the reproductive pipeline at any one time: a dormant or diapausing blastocyst, a small joey nursing and developing in the pouch and a larger joey at foot, still suckling.

There has been debate about the adaptive significance of such a pattern. However, the female kangaroo appears to have better control of her reproductive efforts and the resources she puts into them than a comparable placental. The marsupial young in the pouch is fueled only by milk, unlike the placental embryo which is plugged into the maternal blood supply for its nourishment. During poor environmental conditions, such as droughts, the young in the pouch may perish as milk supply is decreased. When this happens a dormant blastocyst resumes development and after the tiny young is born the mother may again mate and produce another blastocyst. The stress is initially greater on larger suckling young in a prolonged drought and the young in the pouch die at progressively earlier ages. In time the female stops breeding altogether.

Through lactation, which may vary with environmental conditions, the female marsupial may maximize her reproductive potential by balancing current success, which may diminish her own immediate survival as well as that of the young, against future reproductive success. Placentals, on the other hand, make extensive investment in their young over a long gestation, during which the chorio-allantoic placenta (partly of paternal origin) alters the females endocrine environment. Whereas selective pressures which favour the maternal genotype continue to maximize her reproductive value, including both present and future success, those selective pressures shaping the father's genes work towards the maximum success of the current attempt at reproduction, the future success of the mother being of less importance. Perhaps the difference between marsupials and placentals is not necessarily one of relative advantage but of a different emphasis on the control of the reproductive effort.

8.5 Evolution of viviparity

As more work is carried out into the comparative physiology of reproduction it is seen that many of the features of mammalian reproduction are in fact phylogenetically old. The hormonal basis of viviparity in mammals was established in the monotremes and probably even very early in vertebrate evolution, even before the appearance of the reptiles. The cleidoic egg of the early reptiles then provided the basis for placentation and intrauterine growth. Perhaps the most significant feature of mammalian reproduction is the evolution of milk feeding. This is at the very base of mammalian evolution. Once lactation was established it opened the door for a variety of mammalian reproductive patterns. Viewed in this light it is not the viviparity of marsupials or even that of placental mammals that is particularly unusual, but lactation and associated maternal care.

9 Energy and Temperature Relationships in Marsupials

9.1 Historical attitudes

Because of early attitudes about the primitive evolutionary status of monotremes and marsupials it is easy to understand the early speculation about their homeothermic abilities. This speculation seemed well founded when early scientists and explorers reported that the body temperatures of monotremes and marsupials were below, and more variable than, those found for placental mammals. The metabolic and thermoregulatory studies by Professor C.J. Martin in 1902 suggested that not only did monotremes and marsupials have lower body temperatures but also very much lower rates of heat production than placentals. From this work the idea became accepted that all anatomically conservative mammals were physiologically primitive. Subsequent studies on groups such as the 'old' placentals from South America, the armadillos and sloths, led to the acceptance of the idea that a low and perhaps variable temperature indicated a primitive or poorly developed level of homeothermism.

The simplicity of the ideas concerning marsupial thermoregulation was not doubted until the 1950s, when new work found that some marsupials had temperature regulating capabilities at least equal to those of placentals. Such a finding would not have been surprising if thought had been given to the extreme environmental conditions in which some marsupials live. For example, the large red kangaroos can withstand the hottest summer days with only the sparse shade of small desert trees for protection. The issue was still clouded, however, by unusual aspects of the marsupial thermoregulatory responses and by their low body temperatures and low levels of basal metabolism. Only recently has the situation been clarified to show that while the marsupials do have such characteristics they are also generally excellent homeotherms.

9.2 Metabolic relationships of marsupials

The levels that have now been established for the basal metabolism (BMR) and resting body temperatures of marsupials are shown in Table 5. When the BMR of the various groups of mammals and birds are compared in weight independent terms, that is, metabolism expressed per $kg^{0.75}$, the average BMR of marsupials is approximately 68% of the average value usually accepted for placentals. This value now established for marsupials is double the value given by C.J. Martin in 1902, but it is still below that of the placentals. As mentioned before, the BMR is considered a good indication of the metabolic abilities of animal groups, in regard to exercise and heat production. On this basis passerine or perching birds have the greater metabolic capability, while

Table 5 Average resting body temperature and basal metabolism of some amniote vertebrates. (After Hulbert, 1980.)

Animal group	Body temperature (°C)	Basal metabolism (Watts per kg$^{0.75}$)
Reptiles		
(*a*) Lizards	30.0	0.40
(*b*) Snakes	30.0	0.48
Monotremes		
(*a*) Echidnas	31.5	0.92
(*b*) Platypus	32.1	2.21
Marsupials	35.5	2.35
Placentals		
(*a*) Edentates	33.0	2.66
(*b*) Advanced	38.0	3.34
Birds		
(*a*) Nonpasserine	39.5	4.02
(*b*) Passerine	40.5	6.92

marsupials should fall well below the placentals in metabolic performance. However, recent work, as we will see, shows that marsupials may have higher capabilities than are suggested by their BMR.

Resting body temperatures of marsupials are also lower than those of the placentals, 2–3°C on the average. It is interesting that each group of animals in Table 5 tends to have a characteristic temperature, and the higher the BMR of the group, the higher the temperature. It has been often suggested that the lower BMR of some groups is related simply to their lower body temperatures. The reasoning was, in the case of mammals, that if the body temperature of monotremes and marsupials were raised to those of placentals, then the Q_{10} effect, the rise in chemical activity due to increasing temperature, would bring the BMR up to a 'proper' mammalian level. This simplistic idea, however, is not supported by the evidence, neither for marsupials nor monotremes.

9.2.1. Metabolic responses to cold

There are two aspects of thermoregulation in homeotherms, the production of sufficient heat to maintain body temperature in cold conditions, and the loss of excess heat when the animal is in hot conditions. In the evolution of homeothermy it has been assumed generally that extra heat production and the maintenance of body temperature in the cold evolved first. This suggestion initially came from the notion that the 'primitive' monotremes had no mechanism for excess heat dissipation. This is not true, but an increased ability to produce heat was at the base of mammalian homeothermy. The first mammals were very small insectivorous types. Their heat producing abilities enabled them to occupy the cool nocturnal insectivorous niche while the reptiles were active during the day. Many small mammals, among them many marsupials, occupy this niche today, and they are still faced with the same basic thermoregulatory problems.

Small homeothermic animals have problems in the cold because they have potentially high rates of heat loss. This is due to their relatively large surface area compared to the volume of heat producing tissue, and the limits on the insulation that a small mammal can carry. The maintenance of body temperature in the face of a high rate of heat loss obviously requires an equivalent level of heat production by small marsupials or any small mammals. Under these circumstances marsupials might have been expected to have thermoregulatory difficulties. Most birds and mammals, in the cold, appear unable to sustain metabolic rates that exceed about four times the BMR. If this was the case with marsupials, their lower basal heat production would mean a relatively limited metabolic capability and restricted cold tolerance compared with that of placentals. Larger marsupials could overcome these problems by increasing insulation, and there is evidence that they do, but are there any alternative options open to small marsupials?

Marsupials are not the same as placentals which have their maximum sustainable (summit) metabolism limited to about four times BMR. Comparisons of the responses of small marsupials and small rodents to very cold temperatures confirmed a summit metabolism of 3–5 times BMR for the rodents but showed that the marsupials were able to increase metabolism 8–9 times so that they had an absolute summit metabolism at least matching those of placentals. The larger expansibility of marsupial metabolism (Fig. 9–1) fits nicely with recent findings on the degree of flexibility in the cardiovascular and respiratory systems of marsupials.

9.2.2 *Torpor in marsupials*

Insight into other aspects of marsupial thermoregulation, such as their use of torpor, has increased recently. Torpor, particularly short term or daily torpor, is widespread among small marsupials of many families. It is apparently used for energy conservation, since food restriction is often necessary to bring torpor about in laboratory conditions. Torpor is not a switching off of the thermoregulation system, as was thought some years ago, but is a well controlled thermoregulatory process. Many of the small marsupials only allow body temperature to drop to 15–17°C, even if air temperature falls lower, and a larger drop in air temperature will initiate arousal. Some marsupials do show a pattern of longer term torpor much like the seasonal hibernation of placentals. The pigmy possums (Family Burramyidae) show this pattern and allow their body temperatures to drop to much lower levels.

One aspect in which marsupial torpor differs from that of placentals is in the characteristics of arousal. Placentals rely on brown fat, a special heat producing tissue to accelerate rewarming on arousal. Marsupials do not appear to have brown fat and the mechanism of their arousal is as yet not understood.

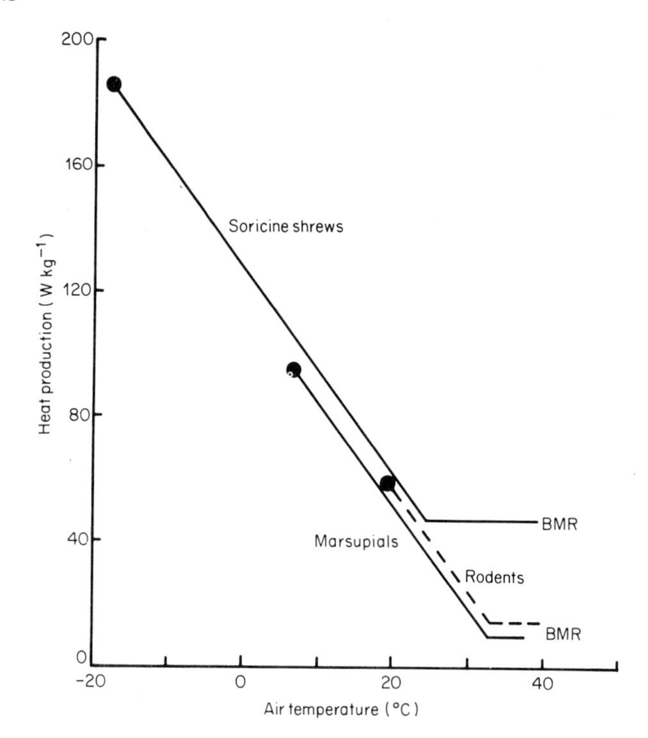

Fig. 9–1 Pattern of heat production in response to cold of small marsupials and placentals. The circles indicate maximum sustainable (summit) metabolism. Marsupials have a high summit metabolism and metabolic scope compared with most placentals, e.g. rodents. Small soricine shrews have elevated total metabolism to obtain a high summit metabolism and resistance to cold.

9.3 Energetics of locomotion in marsupials

The metabolic response to exercise by marsupials is of interest particularly because of the hopping locomotion of the kangaroos, these being the largest animals to hop. With quadrupeds generally there is a relatively constant increase in the cost, or power requirement of locomotion as speed increases. Quadrupedal marsupials also tend to follow this pattern, but the hoppers do not (Fig. 9–2). During hopping the change in power requirement with the changing speed follows an unsual pattern, so that a kangaroo travelling at moderate speeds does so more economically than do similarly sized running bipeds or quadrupeds.

If hopping has advantages, why is it so rare in large animals? The answer may be found by examining the locomotion pattern of kangaroos at low speeds. At

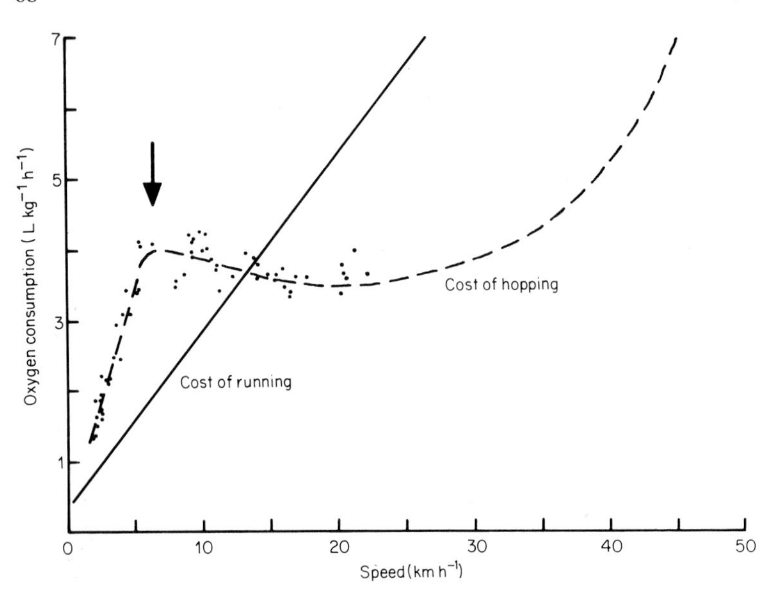

Fig. 9–2 The energetic cost of hopping in kangaroos as compared with the cost of running in quadrupeds and bipeds. Cost is given as oxygen consumption. Arrow indicates the point of transition from pentapedal locomotion to hopping. The solid line indicates the generalized cost of running for an animal of the same size as the kangaroos (18 kg). (After Dawson, 1977.)

speeds below 6 km h^{-1}, kangaroos do not hop. They move in a rather odd way, using their heavy tail as an additional support. This gait has been called pentapedal because the tail acts as a fifth leg to help the forelimbs support the animal as the large hind limbs are moved forward together. (Contrary to a widespread belief, kangaroos can move their hind legs independently. They are good swimmers, and as they swim their hind legs alternate.)

This pentapedal gait is clumsy and energetically costly. The reason kangaroos resort to this odd mode of locomotion at low speeds is that it would be even more costly in terms of energy to hop. A simple physical analysis of the various energetic components of hopping, particularly those relating to work against gravity supports this view. The same kinds of factors explain why it is relatively costlier for large animals to climb trees or run up slopes than it is for small animals. Small hopping animals use quadrapedal locomotion at low speeds, but kangaroos seem precluded from this by the gross anatomical specialization of their hind limbs and the reduction in size of their forelimbs. These specializations are related to increased body size and some smaller macropodids do use quadrupedal locomotion at low speeds.

At low speeds large kangaroos have problems, but at high speeds hopping has advantages over running. Once hopping starts, the energetic costs do not change over a wide range of speeds. In treadmill experiments, when kangaroos hopped at speeds above 15 km h^{-1}, they travelled more economically than running animals. This pattern of energy expenditure appears to be explained by the combination of a fixed hopping rate and the storage of energy in elastic elements such as tendons. Energy can be stored in such elastic fibrous tissues much as it is stored in the spring of a pogo stick.

Treadmill measurements are not practical at speeds above 22 to 23 km h^{-1}, but some overall comments can be made on the basis of observations of small hopping animals and field observations of kangaroos. It may be assumed that animals normally travel at their most comfortable speed, or economical speed. For kangaroos this appears to be between 20 and 25 km h^{-1}. Energy costs probably increase above this speed because the elastic storage of energy cannot increase indefinitely. Hopping rate or stride frequency, however, remains constant up to about 40 km h^{-1}, with the increase in speed being achieved by increasing the length of the stride. If a kangaroo is pressed, it can maintain this speed for a distance of a couple of kilometres. Kangaroos can even increase their speed above 40 km h^{-1}, but they rarely do so. They hop in such high-speed bursts only in emergencies, and then only for a few hundred metres. Kangaroos travelling at speeds of up to approximately 50 km h^{-1} are often seen in the field, and speeds of 65–70 km h^{-1} have been reported. At these speeds there is a noticeable increase in stride frequency as well as stride length. The increase in stride frequency would be accompanied by a marked increase in energy consumption, but exactly how the energetic cost of hopping at maximum speeds is related to the cost of high-speed quadrupedal locomotion is unknown.

It was once considered that marsupials, with their low BMR, may be restricted in metabolic capability, and hence maximum energetic performance. In this light, the hopping of kangaroos might reflect a way around this limitation. However, as was noted for the summit metabolism, this does not now seem to be the case, at least for the Australian species. Kangaroos increase their aerobic metabolism by more than 20 times during hopping, indicating that marsupials may have higher metabolic capabilities than placentals.

Why did hopping come into existence? Here two aspects of locomotion must be considered. One is the need of maximum sustained speed for escaping danger. The other is the need for economy in long distance travel. It may be that hopping was developed to take advantage of the apparent high metabolic scope under one of these conditions. However, the answer will have to wait for further work on the energetics of hopping. Hopping as a mode of locomotion must have advantages in spite of its shortcomings at low speed. Kangaroos and other macropodids have radiated widely throughout Australia and many species have survived the coming not only of aboriginal man and his fellow traveller, the dingo, but also of European man.

9.4 Responses of marsupials to heat

Early thermoregulatory studies on marsupials concluded that marsupials were deficient in their ability to maintain body temperature in hot conditions, and that they had only a limited capacity for evaporative cooling. The techniques used in these studies were poor and marsupials are now known to have excellent heat dissipating systems, comparable at least with those seen in placentals. Panting, sweating and licking may be used, but panting is the most widely used form of evaporative heat loss among marsupials. As in the case of most placentals, it takes the form of a graded response in respiratory rate. Unlike the panting seen in some placentals, such as the dog, the air is passed in and out through the nose rather than the mouth. In kangaroos there are elaborate arrangements of the nasal turbinates which bring air close to the nasal lining, presumably to help evaporation.

Sweating seems to be important only in the macropodids where it is used largely as an auxillary mechanism to cope with the high heat production associated with exercise. The method of sweating in the macropodids is in some cases unusual. The rat kangaroo, *Potorous tridactylus*, which is the smallest mammal that sweats, weighs only about 1 kg, but sweats profusely from its naked tail. The body fur is long, and provides insulation when the rat-kangaroo is at rest in cool conditions, but during exercise and hot conditions this active little marsupial evidently has to have an extra route of heat loss in addition to its respiratory evaporation.

Licking, the spreading of saliva and its evaporation, is utilized as a heat loss mechanism by many mammals, both large and small. It is generally used as an auxillary mechanism and is often seen in mammals with limited abilities to cope with overheating, particularly small ones such as the rodents, which normally avoid high temperatures. It has been regarded as inefficient and primitive. The use of licking by some marsupials, such as opossums and kangaroos, initially was thought to support this assumption. Kangaroos do lick their forelimbs during heat stress, but they also pant and sweat. The full interrelation of these different methods of evaporative heat dissipation is seen only during exercise, however. When a kangaroo is resting, it dissipates heat primarily by panting; it does not sweat, but in severe heat it may also lick its forelimbs.

The kangaroo's licking behaviour has been a puzzle because the area of the forelimb which is usually licked is small (Fig. 9–3) and it seemed that the overall benefit in heat dissipation would be limited. The blood vessels in the forelimbs of kangaroos however, form a dense, superficial network in the region the animals usually lick, and during heat stress the blood flow to this region is greatly increased. The forelimb region can thus be a site of significant heat transfer. Indeed, by spreading saliva on their forelimbs kangaroos may well be making the most efficient use possible of an overflow of fluid from their respiratory system, the principal site of evaporative heat dissipation in resting animals.

9.5 Heat loss in exercising kangaroos

The pattern of heat loss in exercising kangaroos is very different from that generally seen in mammals. During exercise in kangaroos sweating replaces

Fig. 9–3 Posture of a desert kangaroo in sparse shade on a hot summer day. Tail is pulled between the legs to reduce the surface for radiant heat inflow. Licking of the vascularized region of the forearms helps heat loss, but panting is still the major form of heat loss.

panting because shallow panting is not possible when the animal is exercising. Evaporation from the respiratory system is still very important because when the animal is exercising, water loss from the respiratory surfaces is greatly increased, and is greater than the water loss by panting when the animal is at rest. This increased respiratory ventilation is in response to the increased demand for oxygen during exercise and is therefore limited by the level of oxygen intake. Body sweating enables the animal to bring into play an additional surface area for further evaporative heat loss if heat production outstrips respiratory dissipation. Licking is not possible when an animal is hopping, although hopping kangaroos do stop occasionally and spread saliva on their forelimbs. Licking after exercise seems to assist the rapid return to normal body temperature if body temperature has become elevated by exercise.

Another aspect of their thermoregulatory response to exercise that sets kangaroos apart from other mammals, is that sweating stops as soon as a kangaroo stops exercising, even when its body temperature is elevated and the animal is still panting and vigorously licking its forelimbs. Other mammals capable of sustained exercise, such as horses and men, do sweat to dissipate heat when they are exercising but also sweat to dissipate heat when they are at rest. A few mammals, such as cattle, sheep and other bovids, sweat and pant simultaneously.

What does the kangaroo achieve by ceasing to sweat and changing to panting once it stops exercising? The answer lies in water economy. An animal that lives in an arid environment must conserve its body water, and water is more efficiently used for thermoregulation when it is evaporated from the respiratory tract rather than from the skin. Evaporation from the skin results in a lowered surface temperature and markedly changed gradients for heat flow. In hot conditions this leads to a high flow of heat from the environment and extra water is required to eliminate this additional heat. Panting does not lower the body surface temperature.

On a very hot summer day, the red kangaroo of the arid regions of Australia is an example of an animal doing the most it can to regulate its body temperature and conserve its water. When the radiation heat load is high and air temperature is higher than 45°C, the red kangaroo does not lie down in the shade but stands hunched, thus presenting the smallest amount of surface area for the uptake of heat, particularly that in the form of solar radiation. In this role the animal's dense fur also provides an almost ideal shield against the environment's heat. To further minimize the surface area exposed to solar radiation the kangaroo's long, thick tail is pulled between its legs (Fig. 9–3). The tail too has a complex network of superficial blood vessels that are involved in heat dissipation, but when conditions are severe and favour heat flow into the body, this system is closed down.

The great efficiency of the mechanisms used by arid zone kangaroos to save water in hot dry conditions is shown by the relative amounts of water used by kangaroos and placentals (Table 6). The large differences between the marsupial kangaroos and placentals reflects not only the efficiency of water conservation in temperature regulation in the kangaroos, but also many behavioural adjustments. Such adjustments relate to micro-environment selection, time of activity and feed selection.

Oddly, as more work is carried out on the energetic and thermoregulatory characteristics of marsupials, it becomes apparent that their 'primitiveness' is not real. On the contrary, they have abilities which are in many respects superior to those of the placentals.

Table 6 Water use and total body water during summer of free ranging marsupials and placentals in the arid zone. (After Dawson, Denny, Russell and Ellis, 1975.)

Species	Weight (kg)	Water use (ml kg^{-1} day^{-1})	Total body water (% body weight)
Red kangaroo	22	40	69
Euro	28	39	73
Goat	28	115	67
Sheep	28	173	66

10 Aspects of Marsupial Cardiovascular and Respiratory Function

Considering the differences in aerobic metabolism between marsupials and placentals, other functional differences might be expected in the physiology of the two groups of mammals. In particular, the cardiovascular and respiratory systems, which are concerned with oxygen transport, may differ between the two groups. Some early work was contradictory in this regard but later evidence, both direct and indirect pointed to different functional relationships in the cardiovascular and respiratory systems of the two groups. Notably, the resting heart rates of marsupials were found to be less than half those of placentals.

10.1 Allometric relationships

Difficulties in making comparisons between animal groups arise because of differences in body size. Many anatomical and physiological variables do not change directly in proportion to changes in body mass. Often a non-linear relationship exists between a variable and body mass and these are usually described by allometric equations. Such equations have the form $y = aM^b$, where y is a physiological or anatomical variable, which is a function of body mass (M in kg) raised to an exponent b. These equations are called allometric equations because the relationships do not follow precise geometric scaling. In the logarithmic form these equations give a linear function, $\log y = \log a + b \log M$. The constant a (y-intercept) appears characteristic for different phylogenetic groups, while b, the mass exponent, or the slope of the line in a logarithmic graph, appears constant for a particular variable, even between animal groups. Such an example is the exponent of 0.75, which is associated with the allometric equations for basal metabolism.

10.2 Cardiovascular and respiratory allometry in marsupials

When allometric equations for the resting heart rates of marsupials and placentals were initially compared, they had similar exponents, -0.26, but the levels were very different and the y-intercept in marsupials was less than half that in placentals. It was originally suggested that such a difference could be due to a large difference in metabolic activity or to marked differences in cardiovascular function. The low metabolic rate later found for marsupials did not fully account for the low resting heart rates of the marsupials, and this prompted further investigations into the characteristics of the cardiovascular and respiratory systems of marsupials. The results of these studies are summarized in Table 7 and take the form of a series of allometric equations for different cardiovascular and respiratory variables.

Table 7 Comparison of allometric equations relating cardio-respiratory variables to body mass in resting marsupials and placental. (Data from Dawson and Needham (1981), + except for heart mass which are later results.)

Variable (units)	Marsupial	Placental	Marsupial/Placental
$\dot{V}_{O_2}$ (ml min^{-1})			
(a)*	$7.1M^{0.75}$	$10.0M^{0.75}$	0.71
(b)*	$7.5M^{0.72}$	$11.6M^{0.76}$	$0.65M^{-0.04}$
f_h (min^{-1})	$116M^{-0.24}$	$241M^{-0.25}$	$0.48M^{0.01}$
$\dot{V}_s$ (ml)	$1.22M^{1.07}$	$0.78M^{1.06}$	$1.56M^{-0.01}$
$\dot{V}_b$ (ml min^{-1})	$135M^{0.87}$	$187M^{0.81}$	$0.72M^{0.06}$
Heart mass+ (g)	$7.7M^{0.98}$	$5.8M^{0.98}$	1.33
f_r (min^{-1})	$22.1M^{-0.26}$	$53.5M^{-0.26}$	0.41
$\dot{V}_T$ (ml)	$12.1M^{0.98}$	$7.7M^{1.04}$	$1.57M^{-0.05}$
$\dot{V}_E$ (ml min^{-1})	$267M^{0.72}$	$379M^{0.80}$	$0.72M^{-0.08}$

(a)* normally accepted BMR values
(b)* minimum oxygen consumption values during experiments in which the cardiovascular measurements were made.

Abbreviations: $\dot{V}_{O_2}$ oxygen consumption; f_h, heart rate; $\dot{V}_s$, heart stroke volume; $\dot{V}_b$, cardiac output; f_r, respiratory rate; $\dot{V}_T$, respiratory tidal volume; $\dot{V}_E$, respiratory minute volume.

Most of the cardiovascular and respiratory variables of marsupials can be seen to change allometrically with body mass; with the exponents, *b*, similar to those for placentals and birds. However, the *y*-intercept values for the marsupials differ from the values compiled for placentals. The equations point to very marked differences and indicate that many of the cardiovascular and respiratory characteristics of marsupials are substantially different from those of placentals.

The relationships of the cardio-respiratory system of resting marsupials appear to differ from those of placentals in two fundamental ways (Table 7). Basic phases of the transport of oxygen, initially from the environment to the lungs as measured by respiratory minute volume ($\dot{V}_E$), and then from the lungs to the tissues as measured by cardiac output ($\dot{V}_b$) seem adjusted in accordance with the difference in the standard oxygen consumption between the two groups. Both in the case of $\dot{V}_E$ and of $\dot{V}_b$ the exponents of mass were close to 0.75, that for basal metabolic rate. The exponents for $\dot{V}_b$ are a little higher but this may be related to a small decrease in oxygen extraction with increasing size. The second aspect of the difference in the cardio-respiratory systems of marsupials and placentals is the much lower resting heart and respiratory rates of marsupials, lower than would be anticipated simply from the differences in minimal oxygen consumption. Associated with these low heart and respiratory rates are the much higher heart stroke volumes and lung tidal volumes than seen in placentals. The hearts of marsupials are also relatively larger than those of placentals, reflecting the difference in stroke volumes.

Although the differences in the cardio-respiratory systems of marsupials and placentals appear large, there are certain aspects which point to a similarity in 'engineering'. This is shown by ratios between several of the cardio-respiratory variables (Table 8). The residual mass indices resulting from the calculation of these ratios are negligible and so the ratios can be regarded to be mass independent. These ratios such as $\dot{V}_{O_2}$: $\dot{V}_E$, the oxygen extraction ratio and $\dot{V}_E$:$\dot{V}_b$, the ventilation-perfusion ratio are given for the two groups of mammals and also for birds. Between the mammals the similarity is marked, note the O_2 extraction ratio and the ventilation-perfusion ratio. Such a situation probably reflects the basic similarities of their cardio-pulmonary systems as well as their close evolutionary affinities. The volume ratios, $\dot{V}_T$:$\dot{V}_s$, and the frequency ratios, f_r:f_h, are also similar in marsupials and placentals. The differences in the cardio-respiratory systems of these mammal groups appear to arise because the proportionality or balance between volumes and frequencies differ in each group. Birds, with their more complex respiratory system, exhibit a different pattern which is characterized by their higher oxygen extraction ratio.

The overall implications to marsupials of the different balance in their cardio-respiratory system could have been difficult to assess, since the allometric equations in Table 7 apply only to resting conditions, if it was not for other evidence concerning the metabolic and cardio-respiratory scope of these animals. The maximum heart rates of marsupials and placentals have been found to be similar during exercise. Since heart stroke volume remains constant during exercise, marsupials with their larger $\dot{V}_s$ probably have a greater cardiac output capability. This seems surprising for a group of mammals often regarded as primitive in their metabolic and thermoregulatory responses. However, as noted in Chapter 9, marsupials have a high summit metabolism, 8–9 times the BMR, compared with 3–5 times the BMR for placentals. In addition to the high summit metabolism noted for marsupials, hopping kangaroos can also sustain an oxygen consumption at least 20 times basal levels. This information supports the conclusion that the cardiovascular and respiratory systems of marsupials are very expandable and have a high capacity.

Table 8 Mass independent ratios between some basic cardio-respiratory variables in marsupials, placentals and birds. (Data from Dawson and Needham, 1981.)

Ratio	Marsupial	Placental	Bird
$\dot{V}_{O_2}$:$\dot{V}_E$	0.027	0.026	0.040
$\dot{V}_E$:$\dot{V}_b$	1.98	2.03	1.72
$\dot{V}_T$:$\dot{V}_s$	9.9	9.9	12.5
f_r:f_h	0.21	0.25	0.11

11 Brain and Intelligence

The most significant impression formed from an examination of a series of marsupial brains is their fundamental similarity to the brains of placental mammals. Given their independent evolutionary histories this is surprising and demonstrates a remarkable degree of evolutionary parallelism. In several cases similar neural solutions seem to have occurred in marsupials and placentals, and this presents the possibility for investigating the factors influencing brain evolution. Before discussing these aspects however, it is necessary to consider the general features of the mammalian brain and also the accepted notion of the primitive nature of the marsupial brain and associated limited intelligence.

11.1 Basic brain structure

The central nervous system of vertebrates can be considered as a tube of nervous tissue, the spinal cord, which runs the length of the body. Nerve fibres carrying information to and from locations in the body enter and leave the cord along its length. The fibres carrying incoming (afferent) sensory information enter the cord on its dorsolateral aspects in a series of fibre bundles, the dorsal roots, and their cell bodies generally lie outside the cord. Fibres carrying outgoing (efferent) information also leave the cord as a series of bundles, this time on the ventrolateral aspect, and are called the ventral roots. These motor fibres have their cell bodies located ventrally within the cord and they terminate directly in voluntary muscle or on other neurons which in turn communicate to smooth muscle or secretory glands. This segregation within the spinal cord results in a dorsal pair of sensory cell masses, and a ventral pair of motor cell masses. Connecting these four cell masses is the internuncial cell mass (Fig. 11–1).

In marsupials as in other higher vertebrates, the front of the spinal cord is greatly expanded into a set of enlarged vesicles surrounded by specialized masses of brain tissue. The three major subdivisions are the forebrain, midbrain and hindbrain (Fig. 11–1). A pair of large outgrowths, the cerebral hemispheres, occur on either side of the forebrain, while a third major outgrowth, the cerebellum, lies over the roof of the hindbrain. All of these brain enlargements are associated with the growth of large groups of sensory centres which receive the input from the specialized receptors concentrated in the head region (Fig. 11–1). The olfactory bulbs are the foremost of these sensory receptors and are outgrowths of the forebrain. A primary visual input terminal occupies the anterior tectum on the roof of the midbrain. The posterior tectum and associated hindbrain regions receive auditory input from the ears, while areas below the cerebellum receive input from the vestibular

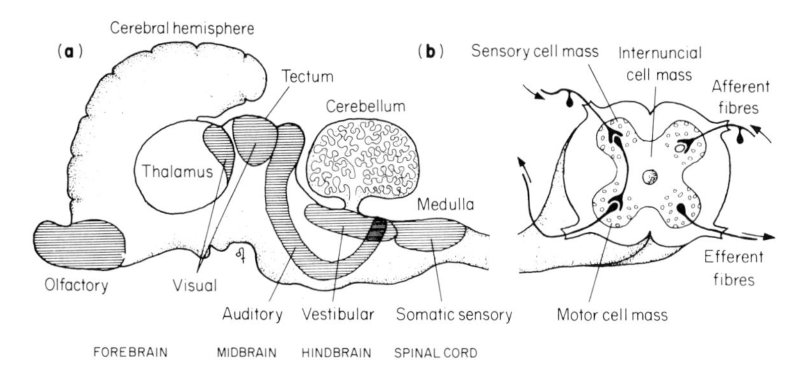

Fig. 11–1 (**a**) A section of the brain of the virginia opossum showing the major regions of brain. Shaded sensory areas are regions which receive the inputs from the specialized receptors in the head region. (**b**) Basic organization of the cell masses in the spinal cord of the virginia opossum. Also shown are the paths of afferent and efferent nerve bundles. (After Johnson, 1977.)

and labyrinthine sensors of gravity and motion, also in the ears. Body sensory inputs via the spinal cord also accumulate in the hindbrain behind the cerebellum and this region together with that dealing with vestibular inputs provide major connections to the cerebellum.

These sensory regions generally send information to a large cell group at the centre of the forebrain, the thalamus. The enlarged cerebral hemispheres have developed primarily as a processing system, handling and organizing this sensory input to the thalamus. The lateral and dorsal surfaces of the paired hemispheres of the forebrain are covered by the pallium or cerebral cortex. In mammals this falls into three main divisions. The central division, neocortex, is surrounded by the paleocortex and the archicortex. The paleocortex is basically concerned with olfactory inputs while the archicortex (hippocampal formation) seems to be involved in processing and passing on information from the midbrain to the forebrain.

The neocortex with its closely associated dorsal thalamus is the most distinctive feature of mammalian brains when compared with those of other vertebrates. It is the expansion and ramification of this system which is generally seen as the basis of 'intelligence' and advanced social behaviour. In its least extent, the neocortex is a small cap over the rest of the forebrain, in placentals such as hedgehogs and in small marsupial insectivores. It is an extensively developed structure in placentals such as cats and marsupials like wombats and kangaroos. However, extreme neocortical predominance as seen in primates, cetaceans and monotremes is not found in marsupials.

11.2 Brain size and intelligence

There is a pervading view that marsupials generally have less developed cerebral cortical regions of the brain and are inferior to placentals in intelligence and sociability. In its simplest form this argument deals only with relative brain size, and extrapolates to relative intelligence. The argument about the brain size and functional capacity is based on the simple proposition that if the size of the smallest components of the brain, the neurons, is relatively fixed, then the functional capacity of the system will depend upon its total population of neurons, as shown by absolute size, and on its structural differentiation. In both marsupials and placentals it is the development and proliferation of the neocortex and the closely associated dorsal thalamus that distinguishes them from the other vertebrates. Therefore, it is the extensive development of these areas which is largely associated with increases in brain size.

The general argument about brain size and intelligence arose last century, but its best known proponent is H.J. Jerison, who has developed an encephalization quotient, EQ, for various groups of vertebrates. He has achieved this by first empirically deriving the allometric relationship between brain mass and body mass in the particular groups. The equation used has the form $E = kM^{0.67}$: where E is brain mass in g, M is body mass in kg, and k is the proportionality constant. The mass exponent 0.67 is close, but not equal to the actually derived exponent. Jerison (1973) used 0.67 because 'an exponent of 0.67 implies a surface:volume relationship and may therefore be the basis for theorizing on the significance of brain size'. This reasoning is not sound but the potential bias is not large. Other workers have used the actual exponents and these are generally lower, 0.56–0.64. The encephalization quotient is the ratio of actual measured brain size to the expected brain size, the latter being calculated from Jerison's allometric relationship. For all living mammals, including the higher primates and cetaceans, the expected brain size, Ee, is given as $Ee = 0.12M^{0.67}$.

Didelphis marsupialis has an EQ of 0.22, which is comparable to those of mesozoic mammals. This value has been used to place marsupials generally in a category of primitive, small brained and unintelligent mammals. Some of the placental insectivores are also in this category. The true picture with regard to marsupial brains and the adaptive radiation of marsupials appears somewhat different, however. It has been apparent for sometime that an assessment of marsupials based solely on *Didelphis* is false, because there are other marsupials with relatively much larger brains. An upward trend of brain mass relative to body mass, associated with an expanison of the neocortex is now known to exist, increasing from the didelphids to the dasyurids, phalangerids and macropodids, with the macropodid level equivalent to the middle of the placental range.

A recent more detailed study of Australian marsupials by Nelson and Stephan (1982) has added much new data and clarified the picture. Using extensive data on the Dasyuridae as their baseline (with a value of 100) they derived encephalization indices for 89 species of Australian marsupials. The value 100 is approximately half the predicted brain mass calculated by Jerison for living placental mammals but his calculation, it should be remembered,

Table 9 Encephalization indices of some marsupials using Dasyuridae (100) as a base. (After Nelson and Stephan, 1982.)

Group	Encephalization index	Comments
Didelphidae	116 (75–137)	*Didelphis* spp. 75
Dasyuridae	100 (85–120)	Planigale species (39–61) omitted
Thylacinidae	148	
Peramelidae	99 (84–122)	
Thylacomyidae	128	*Macrotis* only
Phalangeridae	130 (120–140)	
Burramyidae	98 (89–105)	
Petauridae		
Petaurus-group	156 (139–171)	Including *Gymnobelideus* sp.
Pseudocheirus-group	83 (69–105)	Including *Schoinobates* sp.
Dactylopsila	224	From a single specimen
Macropodidae		
Potoroinae	166 (149–182)	*Hypsiprymnodon* sp. 176
Macropodinae	142 (104–176)	*Petrogale* spp. up to 176

includes the cetaceans and simian primates. The latter includes man with a brain 22 times larger than the Dasyuridae on a mass independent measure. The significant feature of this study (Table 9) is that the dasyurids have indices well above the species of *Didelphis*, but not necessarily above all the Didelphidae.

Many of the diprotodontid marsupials are within the middle of the placental range in brain size, especially the macropodids and the *Petaurus* group of possums. It is interesting that the only value for a marsupial which was generally higher than placentals was *Dactylopsila trivirgata*, the striped possum of northern Australia and New Guinea. This active little possum has been likened to some of the prosimian primates in its behaviour. In searching for food this insectivorous arboreal possum constantly taps its fingers on the bark as if to sound out grubs and termites, which it then extracts from holes and crevices with a long thin specialized finger. Perhaps adaptation to a complex arboreal environment has required an increase in brain complexity and hence brain size in these marsupials.

One aspect of this story is that the cerebral neocortex has a relatively constant thickness in animals of a similar size, as also has the cerebellar cortex. Thus in those species with a large development of the neocortical region of the brain, the increased size is accommodated by an expanded surface area within the restricted volume within the skull. This takes the form of infoldings or convolutions of the cortical surface. As in placentals, the degree of convolutions on the surface of the forebrain of marsupials can be related to body size and 'functional capacity'. Convolution is particularly large in the wombats, large wallabies and kangaroos as well as on the large brains of the Tasmanian devil and the Tasmanian wolf. The amount of neocortex and of cortical folding in these animals is equivalent to that in many placentals.

11.3 Interhemispheric connections in marsupial forebrains

There is another aspect of brain function in which marsupials and placentals differ and in which it has been suggested that marsupials are not as advanced as placentals. This is in regard to the interhemispheric connections in the forebrain. The major distinction between the brains of marsupials and placentals in this regard is in the mode by which the forebrain structures of one cerebral hemisphere communicate with those of the other hemisphere. This difference was recognized in 1837 by Richard Owen. He noted that marsupial mammals lacked a corpus callosum (a major band of fibres connecting the cerebral hemispheres in placentals); he also later noted their absence in monotremes. As only the 'higher' mammals seemed to possess this structure and because it reaches its greatest extent in man, Owen thought that this commissure might be of considerable significance in the evolution of the brain and intelligence.

The primary mammalian pattern of commissural connections is presumably the pattern seen in monotremes and polyprotodont marsupials. In this basic arrangement there are two fibre bundles connecting structures of the two cerebral hemispheres (Fig. 11–2). The more dorsal of these connects the hippocampi of the two sides and is called the hippocampal or dorsal commissure. The more ventral fibre bundle is known as the anterior commissure and carries groups of fibres between the cerebral cortices, in addition to fibres to or from other structures such as the olfactory bulbs. Both the hippocampal and anterior commissures lie in the laminar terminalis, the sheet of tissue forming the front of the third ventricle.

To this basic plan, the diprotodontid marsupials have added another bundle of neocortical commissural fibres (Fig. 11–2), the fasciculus aberrans. The fasciculus aberrans represents a somewhat more direct path from the dorsal neocortex of one side to that of the other. The occurrence of the fasciculus aberrans has also proved to be useful in studies of the phylogenetic relationships within the marsupials. Its absence in the large Tasmanian wolf, *Thylacinus*, demonstrates that it is a diprotodontid characteristic rather than a characteristic of a larger marsupial brain. The fasciculus aberrans serves to resolve relationships between the perameloids and diprotodontids which both have syndactylous feet and between the caenolestoids and diprotodontids, which both have diprotodont teeth. Since neither the perameloids nor the caenolestoids have a fasciculus aberrans they are not closely related to the Diprotodonta.

The corpus callosum of placentals offers an even more direct connection between the hemispheres than the fasciculus aberrans. This is formed when fibres from the cerebral cortex of one side plunge directly across the dorsal aspect of the hippocampus and enter immediately into the cortex of the other

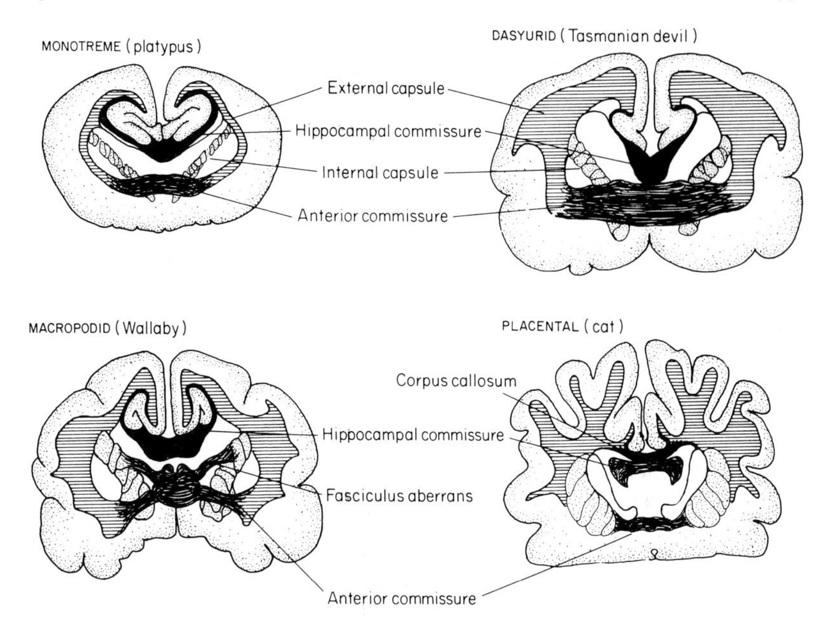

Fig. 11–2 Arrangement of the interhemispheric connections in mammals. (After Johnson, 1977.)

side. The corpus callosum is, as noted, found only in placentals, but its size and shape varies greatly between major placental groups.

Despite the three differing pathways in the different groups of mammals, the fibres connecting the neocortices appear to connect the same elements in much the same way in terms of function and structural considerations, other than route of passage. Fibres connecting the dorsal region of the neocortex pass through the most dorsal commissure available whether it is corpus callosum, fasciculus aberrans or the dorsal portion of the anterior commissure.

Although it has been claimed that the corpus callosum contributed to the evolutionary success of placentals, until the middle of the century little was known of its functions. Part of its function is now known to be the transfer between the hemispheres of visual information gained by either eye. At the optic chiasm there is an initial exchange of visual information before it is passed to the brain. In cats, if the optic chiasm is split, tasks learned using one eye are remembered if only the other eye is used. However, if the corpus callosum is also cut, then this ability to transfer visual information between the hemispheres is lost. It has been suggested that the corpus callosum is an essential part of the development of the interhemispherical transfer and processing of complex information. Some suggest that without a corpus callosum the marsupials may be limited in their ability to process information. However, anatomical studies indicate that in marsupials, the functions of the placental corpus callosum are carried out by other commissures. In the phalangerids visual information is transferred between the hemispheres in a manner equivalent to that in the cat. At this stage the commissure which is involved in this transfer is not known but it seems reasonable to assume that the fasciculus aberrans is involved. The anterior commissure probably plays this role in the other marsupials and monotremes.

11.4 Learning and problem solving in marsupials

Most experimental studies of learning and problem-solving in marsupials have been carried out using the Virginia opossum. The primary interest in these studies was from the point of view that the opossum was a primitive animal, a 'prototype' of placental mammals. From this viewpoint, data on 'the' marsupial was seen to shed light on the development of behaviour patterns and the evolution of intelligence. Preconceptions in experimental studies tend to be self fulfilling, and *Didelphis virginiana* has been reported to be limited in its learning abilities. In recent years further information has become available from studies both on *D. virginiana* and some Australian species, and this has changed the picture somewhat.

Studies to date have included work on discrimination, learning in a variety of forms, maze learning, operant or instrumental learning, and other problem solving and learning tasks such as avoidance and escape learning, and probability learning. The technique most consistently used has been discrimination learning, involving spatial or non-spatial cues. In a spatial discrimination task the animal has to choose between two clearly separated alternatives. A correct response, be it to one side of a T maze or to one of a pair of levers, is rewarded. In spatial tests the positive goal is always in the same position. In non-spatial discrimination tasks the animal has to respond to a specific stimulus which may be moved randomly between trials. Mammals generally learn spatial problems more easily than non-spatial problems.

In discrimination studies with marsupials there has been considerable emphasis on the acquisition of a 'learning set'. Learning set is a characteristic of animals, involving the ability to improve their problem solving capabilities in the process of learning a series of similar problems. This transfer of the ability to solve similar problems has been suggested to be a function of brain complexity. In a variety of experiments it appears that *D. virginiana* develops a learning set only under limited conditions, hence the notion of its low intelligence. However other marsupials, such as the bandicoot (*Isoodon obesulus*) and the brush-tailed possum (*Trichosurus vulpecula*) rapidly acquire a learning set, with *T. vulpecula* at least, being as competent as many placentals. However, assessment of the level of 'intelligence' of marsupials varies noticeably with the type of test. In studies of the ability to solve maze problems *D. virginiana* performed at a level superior to most placental mammals. On the other hand *T. vulpecula* appeared less efficient in maze learning problems than rats and cats.

In other areas of behaviour a dichotomy appears within the marsupials themselves. Didelphid marsupials are suggested to have little development in their play behaviour, supposedly because of their simple behavioural activities. In mammals, two types of play seem to exist, object play and social play. Object, or manipulative, play functions to provide the animal with a diversity of sensory experience and facilitates the acquisition of motor skills. Social play, on the other hand, aids in the acquisition of social skills and presumably has a role in the formation of social bonds. Social play is often considered to be a characteristic of the most advanced of the placentals, yet among the dasyurids social play is very extensive in the juveniles of all the larger species. The larger carnivores of the genus *Dasyurus*, the 'native cats' of Australia, have developed social play to a level on a par with that seen in domestic kittens.

A variety of other types of studies concerned with marsupial behavioural responses have failed to provide a clear assessment of the relative learning capabilities of marsupials and placentals. Even those studies which indicate a superiority of marsupials over placentals in some behavioural aspect need to be viewed with care. Marsupial mice have been reported to be much more reactive to novel situations and habituate much faster to these situations than do placental mice. Such responsiveness to a novel environment may reflect a superior intellectual development since the ability to learn appears related to an animal's exploratory responses to novelty and the rate of habituation to that situation. However, the marsupial mice used in these tests were *Sminthopsis crassicaudata*, aggressive little insectivore-carnivores, whereas the placental mice used in the comparison were normal laboratory bred *Mus musculus*; hardly an appropriate comparison.

It is obvious that broad categorization of marsupials with respect to their learning abilities has to be treated with care. Marsupials as total animals are not homogeneous. They do retain some conservative features, but have many highly evolved derived features. These derived features occur even in areas such as reproduction in which the primitiveness of marsupials is supposed to be demonstrated. In a recent review of the biological strategies of living conservative mammals, Eisenberg (1980) did not even include the diprotodontid marsupials among the six groups of conservative (primitive?) mammals, including four groups of placentals, which he discussed.

Further Reading and References

ALLISON, T. and GOFF, W. R. (1972). Electrophysiological studies of the echidna, *Tachyglossus aculeatus*. III. Sensory and interhemispheric evoked responses. *Archives Italiennes de Biologie*, **110**, 195–216.

ARCHER, M. (1982). A review of the origins and radiations of Australian mammals. Chapter 54 in *Ecological Biogeography of Australia*, ed. A. Keast. Dr. W. Junk Publishers, The Hague.

ARCHER, M. (1982). A review of Miocene thylacinids (Thylacinidae, Marsupialia); the phylogenetic position of the Thylacinidae and the problem of apriorisms in character analysis. Chapter 38 in *Australian Carnivorous Marsupials*, ed. M. Archer. Royal Zoological Society of New South Wales, Mosman, N.S.W.

ARCHER, M., PLANE, M. D., and PLEDGE, N. (1978). Additional evidence for interpreting the Miocene *Obdurodon insignis* Woodburne and Tedford, 1975, to be a fossil platypus (Ornithorhynchidae: Monotremata) and a reconsideration of the status of *Ornithorhynchus agilis* De Vis, 1885. *Australian Zoologist*, **20**, 9–27.

AUGEE, M. L. (1978). Monotremes and the evolution of homeothermy. *Australian Zoologist*, **20**, 111–19.

BAUDINETTE, R. V. (1978). Scaling of heart rate during locomotion in mammals. *Journal of Comparative Physiology*, **127**, 337–42.

BOHRINGER, R. C., and ROWE, M. J. (1977). The organization of the sensory and motor areas of cerebral cortex in the platypus (*Ornithorhynchus anatinus*). *Journal of Comparative Neurology*, **174**, 1–14.

BROWN, T. M. and KRAUS, M. J. (1979). Origin of the tribosphenic molar and metatherian and eutherian dental formulae. Chapter 9 in *Mesozoic Mammals. The First Two-thirds of Mammalian History*, eds J. A. Lillegraven, Z. Kielan-Jaworowska, and W. A. Clemens. University of California Press, Berkeley.

BUCKMANN, O. L. K., and RHODES, J. (1978). Instrumental learning in the echidna. *Australian Zoologist*, **20**, 131–45.

BURRELL, H. (1927). *The Platypus*. Angus and Robertson, Sydney.

CARRICK, F. N. and HUGHES, R. L. (1978). Reproduction in male monotremes. *Australian Zoologist*, **20**, 211–32.

CLEMENS, W.A. (1977). Phylogeny of the Marsupials. Chapter 4 in *The Biology of Marsupials*, eds B. Stonehouse and D. Gilmore. Macmillan Press, London.

CLEMENS, W. A. (1979). Marsupialia. Chapter 11 in *Mesozoic Mammals. The First Two-thirds of Mammalian History*, eds J. A. Lillegraven, Z. Kielan-Jaworowska and W. A. Clemens. University of California Press, Berkeley.

COX, C. B., HEALEY, I. N. and MOORE, P. D. (1976). *Biogeography, An Ecological and Evolutionary Approach*, 2nd edition. Blackwell Scientific Publications, Oxford.

CROMPTON, A. W. (1980). Biology of the earliest mammals. Chapter 1 in *Comparative Physiology: Primitive Mammals*, eds K. Schmidt-Nielsen, L. Bolis and C. R. Taylor. Cambridge University Press, Cambridge.

DAWSON, T. J. (1973). Primitive mammals. Chapter 1 in *Comparative Physiology of Thermoregulation*, Vol. III, ed. G. C. Whittow. Academic Press, New York.

DAWSON, T. J. (1977). Kangaroos. *Scientific American*, **237**(2), 78–89.

DAWSON, T. J. and DAWSON, W. R. (1982). Metabolic scope and conductance in response to cold of some dasyurid marsupials and Australian rodents. *Comparative Biochemistry and Physiology*, **71A**, 59–64.

DAWSON, T. J., DENNY, M. J. S., RUSSELL, E. M., and ELLIS, B. A. (1975). Water usage and diet preferences of free ranging kangaroos, sheep and feral goats in the Australian arid zone during summer. *Journal of Zoology (London)*, **177**, 1–23.

DAWSON, T. J. and FANNING, F. D. (1981). Thermal and energetic problems of semiaquatic mammals: a study of the Australian water rat, including comparisons with the platypus. *Physiological Zoology*, **54**, 285–96.

DAWSON, T. J. and GRANT, T. R. (1980). Metabolic capabilities of monotremes and the evolution of homeothermy. Chapter 13 in *Comparative Physiology: Primitive Mammals*, eds K. Schmidt-Nielsen, L. Bolis, and C. R. Taylor. Cambridge University Press, Cambridge.

DAWSON, T. J. and NEEDHAM, A. D. (1981). Cardiovascular characteristics of two resting marsupials: An insight into the cardio-respiratory allometry of marsupials. *Journal of Comparative Physiology*, **145B**, 95–100.

EISENBERG, J. F. (1980). Biological strategies of living conservative mammals. Chapter 2 in *Comparative Physiology: Primitive Mammals*, eds K. Schmidt-Nielsen, L. Bolis and C. R. Taylor. Cambridge University Press, Cambridge.

GRANT, T. R. and CARRICK, F. N. (1978). Some aspects of the ecology of the platypus, *Ornithorhynchus anatinus*, in the upper Shoalhaven River, New South Wales. *Australian Zoologist*, **20**, 181–99.

GRIFFITHS, M. (1968). *Echidnas*. Pergamon, Oxford.

GRIFFITHS, M. (1978). *The Biology of the Monotremes*. Academic Press, New York.

HARDER, J. D. and FLEMING, M. W. (1981). Estradiol and progesterone profiles indicate a lack of endocrine recognition of pregnancy in the opossum. *Science*, **212**, 1400–2.

HARTMAN, C. (1923). The oestrous cycle in the opossum. *American Journal of Anatomy*, **32**, 353–421.

HUDSON, J. W. and DAWSON, T. J. (1975). Role of sweating from the tail in the thermal balance of the rat kangaroo, *Potorous tridactylus*. *Australian Journal of Zoology*, **23**, 453–61.

HUGHES, R. L. (1977). Egg membranes and ovarian function during pregnancy in monotremes and marsupials. In *Reproduction and Evolution*, eds J. H.

Calaby and C. H. Tyndale-Biscoe. Australian Academy of Science, Canberra.

HUGHES, R. L. and CARRICK, F. N. (1978). Reproduction in female monotremes. *Australian Zoologist*, **20**, 233–53.

HULBERT, A. J. (1980). The evolution of energy metabolism in mammals. Chapter 12 in *Comparative Physiology: Primitive Mammals*, eds K. Schmidt-Nielsen, L. Bolis, C. R. Taylor. Cambridge University Press, Cambridge.

HUNSAKER, D. (1977). Ecology of New World Marsupials. Chapter 3 in *The Biology of Marsupials*, ed. D. Hunsaker. Academic Press, New York.

JERISON, H. J. (1973). *Evolution of the Brain and Intelligence*. Academic Press, New York.

JOHNSON, J. I., Jr. (1977). Central nervous system of marsupials. Chapter 4 in *The Biology of Marsupials*, ed. D. Hunsaker. Academic Press, New York.

KEAST, A. (1977). Historical biogeography of the marsupials. Chapter 5 in *The Biology of Marsupials*, eds B. Stonehouse and D. Gilmore. Macmillan Press, London.

KIRKBY, R. J. (1977). Learning and problem-solving behaviour in marsupials. Chapter 12 in *The Biology of Marsupials*, eds B. Stonehouse and D. Gilmore. Macmillan Press, London.

KIRSCH, J. A. W. (1977a). The classification of marsupials. Chapter 1 in *The Biology of Marsupials*, ed. D. Hunsaker. Academic Press, New York.

KIRSCH, J. A. W. (1977b). The comparative serology of Marsupials, and a classification of Marsupials. *Australian Journal of Zoology, Supplement*, Series No. 52.

KIRSCH, J. A. W. (1977c). The six-percent solution: second thoughts on the adaptedness of the Marsupialia. *American Scientist*, **65**, 276–88.

KIRSCH, J. A. W., and CALABY, J. H. (1977). The species of living marsupials: an annotated list. Chapter 2 in *The Biology of Marsupials*, eds B. Stonehouse and D. Gilmore. Macmillan Press, London.

LILLEGRAVEN, J. A. (1975). Biological considerations of the marsupial-placental dichotomy. *Evolution*, **29**, 707–22.

LYNE, A. G. and HOLLIS, D. E. (1977). The early development of marsupials, with special references to bandicoots. In *Reproduction and Evolution*, eds J. H. Calaby and C. H. Tyndale-Biscoe. Australian Academy of Science, Canberra.

MEYER, J. (1981). A quantitative comparison of the parts of the brains of two Australian marsupials and some eutherian mammals. *Brain, Behaviour and Evolution*, **18**, 60–71.

MURRAY, P. F. (1978). Late Cenozoic monotreme anteaters. *Australian Zoologist*, **20**, 29–55.

NELSON, J. E. and STEPHAN, H. (1982). Encephalization in Australian marsupials. Chapter 58 in *Australian Carnivorous Marsupials*, ed. M. Archer. Royal Zoological Society of New South Wales, Mosman, N.S.W.

PARKER, P. (1977). An evolutionary comparison of placental and marsupial patterns of reproduction. Chapter 16 in *The Biology of Marsupials*, eds B. Stonehouse and D. Gilmore. Macmillan Press, London.

PIRLOT, P. and NELSON, J. (1978). Volumetric analyses of Monotreme brains. *Australian Zoologist*, **20**, 171–9.

RENFREE, M. B. (1980). Placental function and embryonic development in marsupials. Chapter 27 in *Comparative Physiology: Primitive Mammals*, eds K. Schmidt-Nielsen, L. Bolis and C. R. Taylor. Cambridge University Press, Cambridge.

RIDE, W. D. L. (1970). *A Guide to the Native Mammals of Australia*. Oxford University Press, Melbourne.

SHARMAN, G. B. (1970). Reproductive physiology of marsupials. *Science*, **167**, 1221–8.

TYNDALE-BISCOE, H. (1973). *Life of Marsupials*. Edward Arnold, London.

WHITTAKER, R. G., FISHER, W. K. and THOMPSON, E. O. P. (1978). Monotreme haemoglobin and myoglobin amino acid sequences and their use in phylogenetic divergence point estimations. *Australian Zoologist*, **20**, 57–68.

WOOD JONES, F. (1923–25). *The Mammals of South Australia, Parts I–III*. Government Printer, Adelaide.

Index

amniotes 51, 55, 61
Antarctica 8, 33, 36, 37
Archer, M. 48, 49, 50
Asia 36
Australia 7, 8, 11, 25, 28, 33, 35–8 *passim*, 41–7 *passim*, 69, 70, 79
Australian water rat 25, 26
Australo-antarctic continent 36, 37

bandicoots 43, 44, 49, 59, 60, 82
basal metabolic rate *see* BMR
BMR 23
  marsupial, comparison with birds, monotremes, placentals, reptiles 64–6
  monotreme, comparison with marsupials, placentals 23–4
borhyaenids 34, 41–3 *passim*, 48–9 *passim*
brain structure *see* echidnas, mammals, marsupials, monotremes, placentals, platypus
Brown, R. 6
Burramyidae 44–5 *passim*, 79
Burrell, H. 9, 10, 13, 18, 20, 21

Caenolestidae 35, 39, 42–3, 44, 49, 50
Caldwell, W. H. 13
cardio-respiratory physiology
  of marsupials
    allometric relationships 73
    difference from birds and placentals 75
  of monotremes 28
cleidoic eggs 13, 14, 16, 55, 57, 58, 59, 63
climatic changes
  effect on Australian marsupial radiation 37, 38
continental drift 7
corpus callosum 81–2 *passim*
crural system 20–1

Dasyuridae 39, 41, 50, 78, 79
Diatryma 35
Didelphidae 34, 35, 37, 41, 50, 79, 82
Diprotodonta 37, 38, 42, 43, 44–8, 50, 79, 80
diprotodonty 39, 40, 44, 80
duck bill platypus *see* platypus

echidnas *see also Tachyglossus aculeatus, Zaglossus bruijni*
  BMR
    difference from platypus 28–9
  brain structure 29–31
  external features 10, 11, 12
  food 11, 12
  habitat 11, 12
  homeothermy 26, 27
  origins of 8
  long beaked 11–12
  relationship with platypus 6, 7
  reproductive system 13–21
    female 14, 15, 16, 18, 19
    male 15, 19, 20
Eisenberg, J. 83
encephalization quotient 78
EQ *see* encephalization quotient
Eupantotheres 2, 32
Eutheria 1

fasciculus aberrans 80
Fleay, D. 27

Gondwanaland 36
Gregory, W.K. 5, 6
Griffiths, M. 5, 19, 26, 27

hibernation *see* homeothermy
Home, Sir E. 10, 14
homeothermy
  characteristics of 22
  echidnas
    temperature relations of 26
  marsupials
    hibernation 66, 67
    historical attitudes 64
    responses to cold 65–7
    responses to heat 69–72
  monotremes
    hibernation 26–7
    historical attitudes 22–3
    responses to cold 25–7
    responses to heat 27
    differences between monotremes 27
  platypus
    temperature relations of 25–6
    contrast with Australian water